Energy Policies, Politics and Prices

snova
New York

Energy Policies, Politics and Prices

Sustainable and Green Energy Policies
Kürşad Emrah Yıldırım, PhD and Eser Gemici, PhD
2024. ISBN: 979-8-89113-773-8 (Hardcover)
2024. ISBN: 979-8-89113-852-0 (eBook)

The Development of Crude Oil and Natural Gas Prices and the Energy Security of the European Union
Matej Korček, PhD and Saleh Mothana Obadi, PhD
2024. ISBN: 979-8-89113-357-0 (Hardcover)
2024. ISBN: 979-8-89113-591-8 (eBook)

A Strategic Evaluation of Energy Security in the Eastern Mediterranean
Michail Chalaris, PhD
2021. ISBN: 978-1-53619-273-5 (Softcover)
2020. ISBN: 978-1-53619-346-6 (eBook)

The Future of Energy: Challenges, Perspectives, and Solutions
Thomas Valone, PhD (Editor)
2020. ISBN: 978-1-53618-186-9 (Hardcover)
2020. ISBN: 978-1-53618-348-1 (eBook)

Caspian: Status, Challenges, and Prospects. An Analysis into the Legal Classification, Security and Environmental Concerns, Geopolitics and Energy Flow Impact of the Caspian Plateau
Anis H. Bajrektarevic, Mak Bajrekatarevic and Tanvi Kamlesh Chauhan
2020. ISBN: 978-1-53618-956-8 (Hardcover)
2020. ISBN: 978-1-53619-001-4 (eBook)

More information about this series can be found at
https://novapublishers.com/product-category/series/energy-policies-politics-and-prices/

Michael E. Massaro
Editor

Pipeline Safety

Background, Practices and Regulatory Changes

snova
New York

NOTICE TO THE READER

Library of Congress Cataloging-in-Publication Data

ISBN: 979-8-89530-058-9 (Hardcover)
ISBN: 979-8-89530-066-4 (eBook)

Published by Nova Science Publishers, Inc. † New York

Contents

Preface

The U.S. energy pipeline network includes approximately 3.3 million miles of onshore pipeline transporting natural gas, crude oil, and other hazardous liquids. Over the past decade, major safety incidents in California, Massachusetts, Mississippi, and other states have drawn criticism from stakeholders and have raised concerns in Congress about pipeline safety regulation. The 2021 ransomware attack on the Colonial Pipeline has also drawn attention to federal pipeline security activities, including agency roles and the linkage between pipeline safety and security.

Chapter 1 examines the progress of the Pipeline and Hazardous Materials Safety Administration (PHMSA) in implementing the Protecting Our Infrastructure of Pipelines and Enhancing Safety Act of 2020 (P.L. 116–260, PIPES Act of 2020) and examining future needs in pipeline safety. Chapter 2 addresses PHMSA's process to record pipeline safety enforcement information and the extent to which its pipeline enforcement website aligns with key practices for transparently reporting data. Chapter 3 reviews the history and role of the federal program for pipeline safety, including a discussion of pipeline safety trends and major accidents. Chapter 4 identifies key regulatory changes that stakeholders believe will impact gas transmission pipeline safety

Chapter 1

Pipeline Safety: Reviewing Implementation of the Pipes Act of 2020 and Examining Future Safety Needs*

Committee on Transportation and Infrastructure

Committee on Transportation and Infrastructure
U.S. House of Representatives
Washington, DC 20515

Sam Graves
Chairman

Jack Ruddy, Staff Director

Rick Larsen
Ranking Member

Katherine W. Dedrick, Democratic Staff Director

March 3, 2023

I. Summary of Subject Matter

TO: Members, Subcommittee on Railroads, Pipelines, and Hazardous Materials

FROM: Staff, Subcommittee on Railroads, Pipelines, and Hazardous Materials

RE: Subcommittee Hearing on "Pipeline Safety: Reviewing Implementation of the PIPES Act of 2020 and Examining Future Safety Needs."

---

* This is an edited, reformatted and augmented version of Hearing Before the Subcommittee on Railroads, Pipelines, and Hazardous Materials of the Committee on Transportation and Infrastructure, House of Representatives, Publication No. (118–4), dated March 8, 2023.

In: Pipeline Safety: Background, Practices and Regulatory Changes
Editor: Michael E. Massaro
ISBN: 979-8-89530-058-9

II. Purpose

The Subcommittee on Railroads, Pipelines, and Hazardous Materials will meet on March 8, 2023, at 10:00 a.m. ET in 2167 Rayburn House Office Building to hold a hearing titled "Pipeline Safety: Reviewing Implementation of the PIPES Act of 2020 and Examining Future Safety Needs." The purpose of this hearing is to examine the progress of the Pipeline and Hazardous Materials Safety Administration (PHMSA) in implementing the Protecting Our Infrastructure of Pipelines and Enhancing Safety Act of 2020 (P.L. 116–260, PIPES Act of 2020) and examining future needs in pipeline safety.

II. Background

About the Agency

The Pipeline and Hazardous Materials Safety Administration (PHMSA) was created under the Norman Y. Mineta Research and Special Programs Improvement Act of 2004 (P.L. 108–426, "2004 Act"). Prior to enactment of the 2004 Act, the Department of Transportation's (DOT) Research and Special Programs Administration administered the DOT's pipeline and hazardous materials safety programs.[1] PHMSA's mission is to protect people and the environment by advancing the safe transportation of energy through 3.4 million miles of natural gas and hazardous liquid pipelines, which account for the transportation of 65 percent of the energy commodities consumed in the United States. PHMSA also is charged with the safe and secure movement of over one million daily shipments of hazardous materials by all modes of transportation.[2]

The first statute regulating pipeline safety was the Natural Gas Pipeline Safety Act of 1968, which Congress amended in 1976.[3] Congress added hazardous liquid pipelines to the statute in the Pipeline Safety Act of 1970.[4] Recent enacted legislation regulating the safety of natural gas and hazardous

---

[1] Norman Y. Mineta Research and Special Programs Improvement Act of 2004, Pub. L. No. 108–426 [hereinafter the 2004 Act].

[2] See Pub. L. No.117–58; see also PHMSA, Pipeline Safety Program Budget and Grants Presentation (Jan. 25, 2023) (on file with Comm.) [hereinafter PHMSA Budget and Grants Presentation].

[3] Natural Gas Pipeline Safety Act of 1968, Pub. L. No. 90–481 (amended by the Natural Gas Pipeline Safety Act Amendments of 1976, Pub. L. No. 94–477, 90 Stat. 2073).

[4] Pipeline Safety Act of 1970, Pub. L. No. 96–129.

liquid pipeline facilities for which mandates remain outstanding include the Protecting our Infrastructure of Pipelines and Enhancing Safety Act of 2016, and the Protecting our Infrastructure of Pipelines and Enhancing Safety (PIPES) Act of 2020.[5] The current authorization expires at the end of Fiscal Year (FY) 2023 on September 30, 2023.[6]

Pipeline Safety Framework

Safety regulations differ depending on the nature of the pipeline and the commodity that is moving through it. PHMSA's regulations govern pipelines and facilities that transport natural gas separately from those that transport hazardous liquid.[7] Additionally, the pipelines and facilities used to transport natural gas and hazardous liquids vary in operating pressures, diameter size, intended purpose, and proximity to populated areas.[8] This infrastructure includes:

- *Distribution pipelines*: These pipelines transport natural gas to commercial and residential end-users. Gas distribution pipelines tend to be smaller in diameter and operate at lower pressures.[9] PHMSA estimates there are 2.3 million milesof gas distribution lines, much of which are intrastate pipelines.[10] There are no hazardous liquid distribution pipelines.[11]

---

[5] Pipeline Safety Reauthorization Act of 1988, Pub. L. No. 100–561, 102 Stat. 2805; Pipeline Safety Act of 1992, Pub. L. No. 102–508, 106 Stat. 3289; Accountable Pipeline Safety and Partnership Act of 1996, Pub. L. No. 104–304, 110 Stat. 3793; Pipeline Safety Improvement Act of 2002, Pub. L. No. 107–355, 116 Stat. 1757; The 2004 Act; Pipeline Inspection, Protection, Enforcement and Safety Act of 2006, Pub. L. No. 109–468, 120 Stat. 3486; Pipelines Safety, Regu- latory Certainty, and Job Creation Act of 2011, Pub. L. No. 112–90, 125 Stat. 1904; the Protecting our Infrastructure of Pipelines and Enhancing Safety Act of 2016, Pub. L. No. 114–183, 130 Stat. 514; Protecting Our Infrastructure of Pipelines and Enhancing Safety Act of 2020, Pub.L. No. 116–260 [hereinafter PIPES Act of 2020].

[6] PIPES Act of 2020, § 101.

[7] 49 C.F.R. §§ 192, 195 (2023).

[8] Id.

[9] PHMSA, Fact Sheet: Distribution Pipelines, available at https://primis.phmsa.dot.gov/comm/ FactSheets/FSDistributionPipelines.htm (last updated Feb. 26, 2018).

[10] PHMSA Budget and Grants Presentation supra note 2.

[11] GAO, GAO–12–388, PIPELINE SAFETY, COLLECTING DATA AND SHARING INFORMATION ON FEDERALLY UNREGULATED GATHERING PIPELINES COULD HELP ENHANCE SAFETY 3 (2022), available at https://www.gao.gov/ assets/gao-12-388.pdf.

- *Transmission pipelines*: These pipelines transport energy products from treatment and processing facilities to bulk customers, storage facilities, and local distribution networks.[12] The products transported can include natural gas and hazardous liquids[13] Transmission pipelines can range in size from several inches to several feet in diameter and are designed to operate from relatively low pressures to high pressures.[14] These lines can operate within a single State or span hundreds of miles, crossing one or more State lines.[15] PHMSA estimates there are 301,484 miles of gas interstate transmission lines.[16]
- *Gathering lines*: These lines transport natural gas from the production site to a central collection point. PHMSA currently regulates 488,064 miles of gas gathering lines.[17] Historically, gathering lines were built in lower populated areas, had smaller diameters than transmission lines, and operated at pressures and flow lower than transmission lines.[18] However, as new gas development occurs around the country, regulators are considering the impacts of producers building larger diameter and higher pressure gathering lines and a growing national population.[19]
- *Hazardous liquid pipelines*: These pipelines transport liquid petroleum and other types of liquid energy from sources of origin to refineries and chemical plants, and in some cases to storage or

---

[12] PHMSA, *Fact Sheet: Transmission Pipelines, available at* https://primis.phmsa.dot.gov/comm/FactSheets/FSTransmissionPipelines.htm (last updated Jan. 2018).

[13] PHMSA, *Pipeline Miles and Facilities 2010+, available at* https://portal.phmsa.dot.gov/analytics/saw.dll?Portalpages&PortalPath=%2Fshared%2FPDM%20Public%20Website%2Flportal%2FPublic%20Reports&Page=Infrastructure (last updated Jan. 28, 2022) [hereinafter *Pipeline Miles*].

[14] PHMSA, *Fact Sheet: Transmission Pipelines, available at* https://primis.phmsa.dot.gov/comm/FactSheets/FSTransmissionPipelines.htm (last updated Jan. 11, 2018).

[15] PHMSA, *State Programs Overview, available at* https://www.phmsa.dot.gov/working-phmsa/state-programs/state-programs-overview (last updated Sep. 16, 2022) [hereinafter *State Programs Overview*].

[16] *PHMSA Budget and Grants Presentation supra* note 2.

[17] PHMSA, *Safety of Gas Gathering Pipelines* (Jan. 8, 2019), *available at* https://primis.phmsa.dot.gov/meetings/FilGet.mtg?fil=1029.

[18] PHMSA, *Fact Sheet: Gathering Pipelines, available at* https://primis.phmsa.dot.gov/comm/factsheets/fsgatheringpipelines.htm (Jan. 12, 2018).

[19] *See* PHSMA, Notice of Proposed Rulemaking, Pipeline Safety: Safety of Gas Transmission and Gathering Pipelines, 81 Fed. Reg. 29,830 (Apr 8, 2016.) *available at* https://www.govinfo.gov/ content/pkg/FR-2016-05-13/pdf/2016-11240.pdf.

distribution facilities.[20] According to PHMSA, hazardous liquids traverse the United States through approximately 260,000 miles of hazardous liquid pipelines.[21] Hazardous liquids can include energy sources such as crude oil, refined petroleum products, highly volatile liquids, and anhydrous ammonia, and carbon dioxide in the supercritical (fluid or vapor) state.[22] Hazardous liquids can be transported by transmission and gathering lines.[23]

- *Liquefied natural gas (LNG) facilities*: These facilities are used for converting, transporting, or storing LNG. There are several Federal agencies involved in the regulation of LNG.[24] Historically, PHMSA has regulated peak shaving facilities and satellite facilities where LNG has been used to manage capacity during times of peak demand. PHMSA also regulates import terminals.[25] Market changes led to a rapid growth in export terminals; however, PHMSA regulations governing LNG facilities predate such expansion.[26] In 2022, the United States became the world's top LNG exporter tied with Qatar.[27] To address these changes, the PIPES Act of 2016 and the PIPES Act of 2020 mandated that PHMSA update its safety regulations for LNG facilities. PHMSA estimates it will complete this regulation by September 2023.[28]

---

[20] Pipeline Safety Trust, *Hazardous Liquid Pipelines—Basics and Issues, available at* https://pstrust.org/wp-content/uploads/2015/09/2015-PST-Briefing-Paper-03-HazLiquidBasics.pdf (last updated Sept. 2015) [hereinafter *Hazardous Liquid Pipelines—Basics and Issues*].

[21] PHMSA, PHMSA Presentation: Pipeline Safety Program Budget and Grants, (Jan. 25, 2023) (on file with Comm.); see also Pipeline Miles, supra note 13.

[22] Hazardous Liquid Pipelines—Basics and Issues, supra note 20.

[23] Id.

[24] PHMSA, *Jurisdiction of LNG Plants, available at* https://www.phmsa.dot.gov/pipeline/liquified-natural-gas/jurisdiction-lng-plants (last updated Jan. 31, 2018); *see also* 49 C.F.R. § 193.2001 (2023).

[25] *See* PHMSA, *LNG Facility Siting, available at* https://www.phmsa.dot.gov/pipeline/ liquified-natural-gas/lng-facility-siting.

[26] 49 C.F.R. § 193.

[27] Stephen Stapczynski, *U.S. Surges to Top of LNG Exporter Ranks on Breakneck Growth*, BLOOMBERG, (Jan. 2, 2023), *available at* https://www.bloomberg.com/news/articles/2023-01-03/us-surges-to-top-of-lng-exporter-ranks-on-breakneck-growth.

[28] PHMSA, *PIPES Act Web Chart, available at* https://www.phmsa. dot.gov/ legislative-mandates/pipes-act-web-chart (last updated Jan. 24, 2023) [hereinafter *Pipes Act Web Chart*].

PHMSA's Pipeline Safety Oversight

PHMSA sets Federal minimum safety standards for pipeline safety functions, including developing, issuing, and enforcing regulations for the safe transportation of natural gas (including liquefied natural gas) and hazardous liquids by pipeline through the Office of Pipeline Safety (OPS).[29] The Agency's regulatory programs are focused on the design, construction, operation, and maintenance or abandonment of pipeline facilities, and in the construction, operation, and maintenance of LNG facilities.[30] The Agency has jurisdiction over transportation-related facilities; not drilling, siting, or production facilities.[31]

PHMSA has long-experienced difficulty in recruiting and maintaining an inspection workforce capable of meeting PHMSA's oversight needs, as it often competes with industry for personnel.[32] As with previous reauthorization laws, the PIPES Act of 2020 required PHMSA to ensure the number of pipeline inspection and enforcement personnel did not fall below certain levels (224 in FY 21, 235 in FY 22, and 247 in FY 23).[33] It also directed PHMSA to hire eight fulltime employees to finalize outstanding and new congressional mandates, and to seek OPM authority to use recruitment and retention incentives, such as special pay rates, coupled with continued service agreements, that similarly situated agencies have found effective at hiring and retaining staff.[34]

When violations of PHMSA's regulations occur, the Agency has several enforcement mechanisms it can use. These include the issuance of a warning letter, a notice of probable violation, or a corrective action order.[35] The Agency

---

[29] PHMSA, *Office of Pipeline Safety, available at* https://www.phmsa.dot.gov/about-phmsa/offices/office-pipeline-safety (last updated Dec. 13, 2018).

[30] PHMSA, *Pipeline Safety Regulations, available at* https://primis.phmsa.dot.gov/ comm/ SafetyStandards.htm?nocache=8847.

[31] *See* PHMSA, *PHMSA Regulations, available at* https://www.phmsa.dot.gov/regulations (last updated May 5, 2021); *see also* FERC, *Natural Gas Pipelines, available at* https://www.ferc.gov/ industries-data/natural-gas/overview/natural-gas-pipelines (last updated Feb. 10, 2021); *see also* Library of Cong., *Oil and Gas Industry: A Research Guide, available at* https://guides.loc.gov/ oil-and-gas-industry/laws/agencies.

[32] DOT Inspector General, PHMSA has Improved its Workforce Management but Planning, Hiring, and Retention Challenges Remain 2 (2017), *available at* https://www.oig.dot.gov/sites/default/files/PHMSA%20Workforce%20Final%20Report%20 112117-ST2018010.pdf [hereinafter PHMSA WORKFORCE REPORT].

[33] *PIPES Act of 2020, supra* note 5 § 102.

[34] *Id.*; *see also* PHMSA WORKFORCE REPORT, *supra* note 32.

[35] *See* 49 C.F.R. § 190.205 (2023) (allowing for letter notifying an operator of alleged violations and directs them to correct the violation or be subject to additional enforcement action); 49C.F.R. § 190.207 (2023) (providing for a Regional Director to serve a notice—commonly

may also issue fines for non-compliance.[36] In 2022, PHMSA closed 227 enforcement cases.[37]

States' Pipeline Safety Oversight

PHMSA supports states' oversight work by authorizing states to assume certain aspects of pipeline safety for intrastate gas pipelines, hazardous liquid pipelines, and underground natural gas storage through certifications and agreements with PHMSA under 49 U.S.C. §§ 60105 and 60106(a). The Agency also authorizes states with certifications to participate in the oversight of interstate pipeline transportation through agreements under 49 U.S.C. § 60106(b). States with certified pipeline safety programs may impose additional standards for intrastate pipelines and facilities so long as they are compatible with the minimum Federal standards issued by PHMSA.[38]

Cybersecurity

Regarding cybersecurity, PHMSA signed an annex to its Memorandum of Understanding (MOU) with the Transportation Security Administration (TSA) that identifies TSA as the lead entity for pipeline security, which has included cybersecurity threats.[39] In May 2021, the Colonial Pipeline sustained a ransomware attack, causing the operator to shut down the pipeline for six days.[40] PHMSA issued a Notice of Probable Violation (NOPV) and Proposed Compliance Order, and proposed civil penalties of $986,400. The NOPV

---

issued after routine inspections, incident investigations, and other activity—alleging specific regulatory violations and proposing remedial action or civil penalties); 49 C.F.R. § 190.233 (2023) (providing for the Associate Administrator to issue an order finding a particular situation represents a serious hazard to life, property, or the environment and directing certain actions to be taken, up to and including shutdown of the pipeline system).

[36] *See* PHMSA, *Civil Penalty Summary, available at* https://www.phmsa.dot. gov/regulatory-compliance/pipeline/enforcement/civil-penalty-summary (last updated Jan. 25, 2023).

[37] *See* PHMSA, *Listing of Cases Initiated, available at* https://primis.phmsa.dot.gov/comm/reports/enforce/CasesOpenlopidl0.html?nocache=745#lTPl1ltabl2.

[38] State Programs Overview, supra note 16.

[39] PHSMA, Annex to the Memorandum Of Understanding Between the Department of Homeland Security and DOT Concerning Tsa And Phmsa Cooperation on Pipeline Transportation Security and Safety (Feb. 2020), available at https://www.phmsa.dot.gov/sites/phmsa.dot.gov/files/ docs/regulatory-compliance/phmsa-guidance/73466/phmsa-tsa-mou- annexexecuted.pdf.

[40] Vanessa Romo, Panic Drives Gas Shortages After Colonial Pipeline Ransomware Attack, NPR, (May 11, 2021), available at https://www.npr.org/2021/05/11/996044288/panic-drives-gas- shortages-after-colonial-pipeline-ransomware-attack.

alleges that failures to adequately plan and prepare for a manual restart and shutdown operation contributed to the national impacts when the pipeline remained out of service after the May 2021 cyberattack.[41]

III. Pipeline Safety Legislation

In 2021 and 2022, PHMSA finalized seven mandates, including four from the PIPES Act of 2011: Safety of Gas Transmission Pipelines; Discretionary Integrity Management Improvements; Safety of Gas Gathering Pipelines; Amendments to Parts 192 and 195 to require Valve Installation and Minimum Rupture Detection Standards.[42] These final rules completed congressional responses to catastrophic pipeline failures, such as the fatal explosion of a gas transmission pipeline in San Bruno, California in 2010 and the release of one million barrels of crude oil near Marshall, Michigan in 2010.[43] PHMSA also issued an interim final rule on Coastal Ecological Unusually Sensitive Areas, a mandate from the PIPES Act of 2016 andthe PIPES Act of 2020.[44] One mandate from the PIPES Act of 2016 is outstanding: finalizing updates to regulations governing LNG facilities.[45]

---

[41] Letter from Gregory Alan Ochs, Dir., Central Region, Off. of Pipeline Safety, PHMSA to Joseph A. Blount, President and Chief Executive Officer, Colonial Pipeline Company (May 5, 2022) (notifying of probable violation, proposed civil penalty, and proposed compliance order), *available at* https://www.phmsa.dot.gov/ sites/phmsa.dot.gov/files/2022-05/32022026lColoniallPipelinelNOPVlPCP lPCOl05052022.pdf.

[42] Pipeline Safety: Safety of Gas Transmission Pipelines: Repair Criteria, Integrity Manage- ment Improvements, Cathodic Protection, Management of Change, and Other Related Amendments, 87 Fed. Reg. 52,224 (Aug. 24, 2022), *available at* https://www. federalregister.gov/documents/2022/08/24/2022-17031/pipeline-safety-safety-of-gas-transmission-pipelines-repair-criteria-integrity-management; Pipeline Safety: Safety of Gas Gathering Pipelines: Extension of Report- ing Requirements, Regulation of Large, High-Pressure Lines, and Other Related Amendments: Response to a Petition for Reconsideration; Technical Corrections; Issuance of Limited Enforcement Discretion, 87 Fed. Reg. 26,296, *available at* https://www.federalregister. gov/documents/2022/ 05/04/2022-09474/pipeline-safety-safety-of-gas-gathering-pipelines-extension-of-reporting-re-quirements-regulation-of; Pipeline Safety: Requirement of Valve Installation and Minimum Rup- ture Detection Standards, 87 Fed. Reg. 20,940 (Apr. 8, 2022), *available at* https:// www.federalregister.gov/documents/2022/04/08/2022-07133/pipeline-safety-requirement-of-valve-installation-and-minimum-rupture-detection-standards.

[43] Id.

[44] Pipes Act Web Chart, supra note 28.

[45] *PIPES Act of 2016*, Pub. L. No. 114–183, § 27, 130 Stat. 514; *PIPES Act of 2020*, Pub. L. No. 116–260, § 110, 134 Stat. 620; *See* OFF. OF INFO. AND REGULATORY AFFAIRS, PHMSA, DOT, *Pipeline Safety: Amendment to Liquefied Natural Gas facilities* (2018), *available at* https://www.reginfo.gov/public/do/eAgendaViewRule? pubId=201810&RIN=2137-AF45.

In 2020, Congress enacted the PIPES Act of 2020, which directed PHMSA to:

- Determine whether to issue a notice of proposed rulemaking on class location change requirements and to advance the rulemaking process if it makes a positive determination;[46]
- Frequently report publicly on progress made toward completing outstanding mandates;[47]
- Update distribution pipeline safety regulations in response to the September 2018 distribution pipeline explosions in the Merrimack Valley region of Massachusetts;[48]
- Prioritize completion of a final rulemaking on gas gathering lines and direct GAO to study capabilities of mapping gathering lines;[49]
- Strengthen whistleblower protections;[50]
- Update its current regulations for large-scale LNG facilities and allowing it to establish an LNG Center of Excellence;[51]
- Issue regulations prescribing the applicability of pipeline safety requirements to certain idled pipelines;[52]
- Create a technology pilot program, which allows pipeline operators to test and evaluate innovative technologies or practices on their systems that improve pipeline safety;[53]
- Enter into agreement with the National Academies of Science to study potential methods or standards for installing automatic or remote-controlled shut-off valves on pipelines located in certain sensitive areas;[54]
- Issue regulations directing distribution, transmission, and gathering pipeline operators to conduct leak detection and repair programs that protect the environment and pipeline safety, requiring use of advanced leak detection technologies;[55]

---

[46] *PIPES Act of 2020, supra* note 5, § 115.
[47] *Id.* § 106.
[48] *Id.* at Title II.
[49] *Id.* § 112.
[50] *Id.* § 116.
[51] *Id.* §§ 110, 111.
[52] *Id.* § 109.
[53] *Id.* § 104.
[54] *Id.* § 119.
[55] *Id.* §§ 113, 114.

- Require broader transmission of safety-related condition reports to include state authorities, Tribes, and emergency response and planning entities;[56] and
- Prioritize a rulemaking to protect unusually sensitive areas and require operators of certain hazardous liquid pipelines to use internal inspection technology at least once every 12 months and establish procedures for assessing potential impacts by marine vessels.[57]

IV. Future Needs in Pipeline Safety

Pipeline Safety Incidents

PHMSA reports that in 2022, 631 pipeline incidents occurred, from which 10 individuals died, 24 were injured, and property damages totaled more than $685 million.[58] Of these incidents, 17 were classified as serious and 267 were significant.[59]

Carbon Dioxide Pipeline Transportation

There are approximately 5,000 miles of pipeline transporting CO2 primarily for enhanced oil recovery.[60] Carbon capture, utilization, and sequestration (CCUS) is a process that captures CO2 emissions from sources, such as power plants, and either reuses or stores it so it will not enter the atmosphere.[61] The Infrastructure Investment and Jobs Act (P.L. 117–58) establishes a Carbon

---

[56] *Id.* § 121.

[57] *Id.* § 120.

[58] PHMSA, *Pipeline and Hazardous Materials Administration Pipeline Safety Incidents 20 Year Trends, available at* https://portal.phmsa.dot.gov/analytics/ saw.dll?Portalpages&PortalPath=%2Fshared%2FPDM%20Public%20Website%2Flportal%2FSC%20Incident%20Trend&Page=All%20Reported [hereinafter *PHMSA 20 Year Trends*].

[59] *Id.* (explaining Serious incidents involve a fatality or injury requiring in-patient hospitalization and that significant incidents involve a fatality or injury requiring in-patient hospitalization, $50,000 or more in total costs, highly volatile liquid releases of five barrels or more or other liquid release of 50 barrels or more, or liquid releases resulting in an unintentional fire or explosion).

[60] Paul W. Parfomak, CONG. RSCH. SERV., IN11944, Carbon Dioxide Pipelines: Safety Issues, (2022), available at https://crsreports.congress.gov/ product/pdf/IN/IN11944.

[61] DOE, *Carbon Capture, Utilization & Storage, available at* https://www.energy. gov/carbon-capture-utilization-storage.

Dioxide Transportation Infrastructure Finance and Innovation (CIFIA) program for CO2 pipelines and authorizes $2.1 billion over five years for low-interest CIFIA loans and grants through the Department of Energy (DOE).[62]

CO2 can be transported in various states. Currently, nearly all CO2 is transported in a supercritical state (a dense fluid or vapor phase between a liquid and gas) and is regulated under 49 CFR Part 195.[63] Future carbon capture and sequestration (CCS) pipeline projects may be transported in a subcritical state (a liquid) or a gas.[64] PHMSA hazardous liquid regulations address CO2 that is transported as a fluid compressed to a supercritical state and is 90% pure CO2.[65] The potential future use of CO2 with higher levels of impurities which may be transported as a liquid or gas are not regulated by PHMSA.

CO2 is colorless and odorless and in concentrations is an asphyxiant, displacing the oxygen in air and causing negative health impacts.[66] On February 22, 2020, in Satartia, Mississippi, Denbury's 24-inch Delhi Pipeline ruptured, releasing liquid CO2 that immediately began to vaporize.[67] 200 people surrounding the rupture location were evacuated, and 45 people were taken to the hospital.[68] PHMSA's investigation identified contributing factors to the accident, including lack of assessment of geohazards in plans and procedures, underestimating the potential affected areas that could be impacted by a release in its CO2 dispersion model, and not notifying local responders to advise them of a potential failure.[69]

---

[62] *Infrastructure Investment and Jobs Act*, Pub. L. No. 117–58, Division D, 135 Stat. 429, 923.

[63] DOE, Meeting the Dual Challenge, a Roadmap to at-Scale Deployment of Carbon Capture, Use, and Storage 6–4 (2019), *available at* https://www.energy.gov/sites/default/files/2021-06/2019%20-%20Meeting%20the%20Dual%20Challenge%20Vol%20III%20Chapter%206.pdf.

[64] Richard B. Kuprewicz, Accufacts' Perspectives on the State of Federal Carbon Dioxide Transmission Pipeline Safety Regulations as It Relates to Carbon Capture, Utilization, and Sequestration within the U.S., Prepared for Pipeline Safety Trust 4 (2022), *available at* https://pstrust.org/wp-content/uploads/2022/03/3-23-22-Final-Accufacts-CO2-Pipe-line-Report2.pdf[hereinafterACCUFACTS PERSPECTIVES].

[65] 49 C.F.R. § 195.2.

[66] ACCUFACTS PERSPECTIVES, supra note 64.

[67] PHSMA FAILURE INVESTIGATION REPORT, DENBURY GULF COAST PIPELINES, LLC, available at https://www.phmsa.dot.gov/sites/phmsa.dot.gov/files/2022-05/Failure%20Investigation%20Report%20-%20Denbury%20Gulf%20Coast%20Pipeline.pdf.

[68] Id.

[69] Id.

Hydrogen Pipeline Transportation

As of December 2020, there were 1,608 miles of hydrogen pipeline in the United States, primarily on the Gulf Coast.[70] Today, almost all hydrogen is consumed by the oil industry or chemical industry; however, hydrogen pipeline expansion may occur to coincide with emerging hydrogen hubs that offer hydrogen to end users for purposes such as mobility, goods movement, heat for manufacturing processes, and other services.[71] Some United States pipeline operators have initiated projects to blend hydrogen and methane in natural gas pipelines.[72]

V. Witnesses

- Tristan Brown, Deputy Administrator, Pipeline and Hazardous Materials Safety Administration (PHMSA)
- Andrew Black, President and CEO, Liquid Energy Pipeline Association (LEPA)
- Kenneth W. Grubb, Chief Operating Officer—Gas Pipelines, Kinder Morgan, Inc.
- Bill Caram, Executive Director, Pipeline Safety Trust

Wednesday, March 8, 2023

HOUSE OF REPRESENTATIVES,
SUBCOMMITTEE ON RAILROADS, PIPELINES,
AND HAZARDOUS MATERIALS, COMMITTEE ON
TRANSPORTATION AND INFRASTRUCTURE,

---

[70] Paul W. Parfomak, CONG. RSCH. SERV., R46700, Pipeline Transportation of Hydrogen: Regulation, Research, and Policy (2021), available at https://crsreports.congress.gov/product/pdf/R/R46700.

[71] Martin C. OFFUTT, CONG. RSCH. SERV., R47289, Hydrogen Hubs and Demonstrating the Hydrogen Energy Value Chain, (2022), *available at* https://crsreports.congress.gov/product/ pdf/R/R47289.

[72] *Id.*

Washington, DC.

The subcommittee met, pursuant to call, at 10:05 a.m. in room 2167 Rayburn House Office Building, Hon. Troy E. Nehls (Chairman of the subcommittee) presiding.

Mr. NEHLS. The Subcommittee on Railroads, Pipelines, and Hazardous Materials will come to order.

I ask unanimous consent that the chairman be authorized to declare a recess at any time during today's hearing.

Without objection, so ordered.

I also ask unanimous consent that Members not on the sub-committee be permitted to sit with the subcommittee at today's hearing and ask questions.

Without objection, so ordered.

I now recognize myself for an opening statement for 5 minutes.

Opening Statement of Hon. Troy E. Nehls of Texas, Chairman, Subcommittee on Railroads, Pipelines, And Hazardous Materials

Mr. NEHLS. Today's hearing will examine implementation of the Protecting Our Infrastructure of Pipelines and Enhancing Safety Act of 2020, or PIPES 2020 Act, as well as future needs and pipeline safety.

The United States has the largest pipeline network in the world, consisting of about 3.4 million miles of pipeline. The oil and natural gas industry generates about 11.3 million American jobs and accounts for nearly 8 percent of the United States gross domestic product, which is an economic impact of about $1.7 trillion—with a T—per year. The United States is now a leading exporter of energy to the rest of the world, and investments made by the energy sector will continue to provide high-paying jobs for a diversifying workforce and promote domestic energy security.

Pipelines are critical infrastructure for supporting our energy sector and represent the safest, most efficient, most environmentally friendly mode of transporting energy. Pipelines are an essential part of the energy supply chain and ensure communities across the country and our allies abroad have access to affordable and reliable American energy. The recent supply chain crisis and rising inflation, coupled with the war in Ukraine, demonstrate the importance of having reliable, easy access to energy and the negative consequences of restricting or cutting off our energy supply.

The Pipeline and Hazardous Materials Safety Administration, or PHMSA, is the Federal agency responsible for overseeing the safe and reliable operation of the United States pipeline network. Promoting pipeline safety has been a priority for Congress, which enacted several bipartisan laws in recent years to reauthorize crucial pipeline safety programs at PHMSA and direct the agency to initiate actions that advance pipeline safety.

The 2020 pipeline safety reauthorization made advances in safety through some commonsense policy and a balanced, not overly burdensome, regulatory approach. And it is essential that Congress ensure PHMSA is prioritizing and implementing prior congressional mandates and does not impermissibly favor administration priorities over these mandates. PHMSA must prioritize hiring competent pipeline safety inspectors, not more lawyers, and must keep its focus on protecting people and the environment by advancing the safety and reliability of pipelines.

Today, we will hear from PHMSA on progress made towards implementing the PIPES 2020 Act and how it is working to meet the requirements set by Congress. Pipeline safety is a collaborative effort between PHMSA, State governments, industry, and the public. We must continue to promote cooperation between stakeholders and regulators to find a balanced approach that fosters innovation in technology and implementation of best safety practices.

I am looking forward to hearing from industry and public safety representatives today about some of these practices and new technologies that can help us improve safety. This includes how we can support the development of infrastructure to transport new and emerging fuels such as carbon dioxide and hydrogen and how innovative technology and operations are being used to improve inspections, leak detection, leak prevention, and the overall safety of pipeline networks.

As Congress looks to the next pipeline safety reauthorization, we must embrace ways in which the industry and regulators can work together to achieve our goal of zero pipeline incidents. I hope to renew the tradition of passing a bipartisan pipeline safety bill from this committee, and I look forward to working with the ranking member, my colleagues on both sides of the aisle, and the Senate to accomplish this important goal.

[Mr. Nehls' prepared statement follows:]

Prepared Statement of Hon. Troy E. Nehls, a Representative in Congress from the State of Texas, and Chairman, Subcommittee on Railroads, Pipelines, and Hazardous Materials

The United States has the largest pipeline network in the world, consisting of about 3.4 million miles of pipeline. The oil and natural gas industry generates about 11.3 million American jobs and accounts for nearly 8 percent of the United States' gross domestic product, which is an economic impact of about $1.7 trillion per year. The United States is now a leading exporter of energy to the rest of the world. Investments by the energy sector will continue to provide high paying jobs for a diversifying workforce and promote domestic energy security.

Pipelines are critical infrastructure for supporting our energy sector and represent the safest, most efficient, most environmentally friendly mode of transporting energy. Pipelines are an essential part of the energy supply chain and ensure communities across the country, and our allies abroad, have access to affordable and reliable American energy. The recent supply chain crisis and rising inflation, coupled with the war in Ukraine, demonstrate the importance of having reliable, easy access to energy and the negative consequences of restricting or cutting off our energy supply.

The Pipeline and Hazardous Materials Safety Administration (PHMSA) is the Federal agency responsible for overseeing the safe and reliable operation of the United States pipeline network. Promoting pipeline safety has been a priority for Congress, which enacted several bipartisan laws in recent years to reauthorize crucial pipeline safety programs at PHMSA and direct the agency to initiate actions that advance pipeline safety.

The 2020 pipeline safety reauthorization made advances in safety through some commonsense policy and a balanced, not overly burdensome, regulatory approach. It is essential that Congress ensure PHMSA is prioritizing and implementing prior congressional mandates and does not impermissibly favor Administration priorities over these mandates. PHMSA must prioritize hiring competent pipeline safety inspectors, not more lawyers, and must keep its focus on protecting people and the environment by advancing the safety and reliability of pipelines.

Today, we will hear from PHMSA on progress made towards implementing the PIPES 2020 Act and how it is working to meet the requirements set by Congress. Pipeline safety is a collaborative effort between

PHMSA, state governments, industry, and the public. We must continue to promote cooperation between stakeholders and regulators to find a balanced approach that fosters innovation in technology and implementation of best safety practices.

I am looking forward to hearing from industry and public safety representatives today about some of these practices and new technologies that can help us improve safety. This includes how we can support the development of infrastructure to transport new and emerging fuels such as carbon dioxide and hydrogen, and how innovative technology and operations are being used to improve inspections, leak detection, leak prevention, and the overall safety of pipeline networks.

As Congress looks to the next pipeline safety reauthorization, we must embrace ways in which industry and regulators can work together to achieve our goal of zero pipeline incidents. I hope to renew the tradition of passing a bipartisan pipeline safety bill from this Committee, and I look forward to working with the Ranking Member, my colleagues on both sides of the aisle, and the Senate to accomplish this important goal.

Mr. NEHLS. I have a few documents to enter into the record. I ask unanimous consent to enter into the record the following items: a March 8, 2023, letter from the American Public Gas Association; a March 8, 2023, statement from the American Petroleum Institute; and a March 8, 2023, press release from the Distribution Contractors Association and the United Association of Union Plumbers and Pipefitters.

And without objection, so ordered. [The information follows:]

Letter of March 8, 2023, to Hon. Troy E. Nehls, Chairman, and Hon. Donald

M. Payne, Jr.,
Ranking Member,
Subcommittee on Railroads, Pipelines, and Hazardous Materials, from Stuart Saulters,
Vice President of Government Relations, American Public Gas Association, Submitted for the Record by Hon. Troy E. Nehls

MARCH 8, 2023.

The Honorable TROY NEHLS,
Chairman,

Subcommittee on Railroads, Pipelines, and Hazardous Materials, Committee on Transportation and Infrastructure, 2164 Rayburn House Office Building, Washington, DC 20515.

The Honorable DONALD M. PAYNE, JR.,

Ranking Member,

Subcommittee on Railroads, Pipelines, and Hazardous Materials, Committee on Transportation and Infrastructure, 2165 Rayburn House Office Building,

Washington, DC 20515.

Re: Subcommittee on Railroads, Pipelines, and Hazardous Materials (Committee on Transportation and Infrastructure) Hearing: Pipeline Safety: Reviewing Implementation of the PIPES Act of 2020 and Examining Future Safety Needs.

The American Public Gas Association ("APGA") is the trade association representing more than 730 communities across the U.S. that own and operate their retail natural gas distribution entities. These include not-for-profit gas distribution systems owned by municipalities and other local government entities, all accountable to the citizens they serve. As such, APGA's members are critical stakeholders impacted by reauthorization of pipeline safety programs at the Pipeline and Hazardous Materials Safety Administration ("PHMSA") within the United States Department of Transportation ("DOT"). With reauthorization needed in the 118th Congress, we appreciate your holding a hearing on the implementation of the PIPES Act of 2020 and examining future safety needs.

Public gas utilities focus on safe operations, and effective pipeline safety programs implemented by our federal regulator supports our member's ability to deliver energy to Americans. Following are APGA's legislative priorities for consideration during reauthorization discussions.

1. Consider Extending the Natural Gas Distribution Infrastructure Safety and Modernization (''NGDISM'') Grants

The grant program is set to expire after five years, Fiscal Year ("FY") 2026. However, grant award winners can still access awarded money for ten years. This allows reimbursements even after PHMSA stops awarding new funding.

APGA's members have demonstrated tremendous demand for the program. More money was requested in the first funding cycle, about $1.25 billion, than is available for the program in its entirety, $1 billion over 5 years.

PHMSA needed more time than originally anticipated to stand up the program, and as of March 3rd, PHMSA still has not announced the utilities provisionally selected for FY 2022. As well, the grant agreements won't be fully executed until the National Environmental Policy Act ("NEPA") Tier 2 Environmental Review process is completed later in 2023.

PHMSA and Congress seemed to want to target smaller systems with fewer resources through the NGDISM grant program. But many of them were not able to participate in the first funding cycle because they lack the staff and time to devote to grant writing. As well, APGA understands these smaller utilities are interested in learning from successful first round grant recipients to use their resources more efficiently in future funding cycles. The second funding cycle may open before awards are made for the first, which may discourage the smallest systems from applying until year three. Of note, only 179 of the 999 systems eligible applied in 2022.

2. Consider Language Strengthening the Advisory Committee Process

In recent years, PHMSA has occasionally disregarded the recommendations it receives from its Advisory Committees. While it is rare, the impacts are significant. As legislation to reauthorize pipeline safety programs is drafted, APGA would be very interested in engaging with the Committee on Transportation and Infrastructure on establishing a more formal requirement for how often the Advisory Commitees should be meeting and how PHMSA should treat recommendations from these groups.

3. Maintain PHMSA's Cost-Benefit Analysis Requirement

PHMSA must meet a unique cost-benefit analysis requirement before promulgating a new rule. The pipeline safety regulator must ensure its justification for any rulemaking is appropriate prior to sending it to the Office of Management and Budget ("OMB") for review. APGA believes this step results in effective, transparent, and comprehensive regulations and should be maintained.

4. Adequate Funding for Phmsa and Their State Partners

APGA believes that a strong pipeline safety regulator ultimately leads to effective regulations and enhanced pipeline safety. APGA supports PHMSA's full funding and encourages allocation of resources to their state pipeline safety partners. These state partners are responsible for ensuring natural gas distribution operators comply with federal and state, if applicable, pipeline safety regulations.

APGA hopes the above input will be helpful to the Committee. Please do not hesitate to contact me with any additional questions.

Respectfully submitted,
STUART SAULTERS,
Vice President of Government Relations,
American Public Gas Association.

Statement of the American Petroleum Institute, Submitted for the Record by Hon. Troy E. Nehls

On behalf of the American Petroleum Institute (API),[73] we appreciate the opportunity to submit testimony as part of this important hearing addressing pipeline safety, and the reauthorization of the Pipeline and Hazardous Materials Safety Administration (PHMSA) and the pipeline safety program in 2023.

The need for sound energy policy has never been more critical. A modern U.S. infrastructure system, including robust pipeline infrastructure, is crucial to providing a reliable energy supply to every community in America. Every day, a vast network of pipeline infrastructure serves Americans across the

[73] API represents all segments of America's oil and natural gas industry, which supports more than 11 million U.S. jobs and is backed by a growing grassroots movement of millions of Americans. Our nearly 600 members produce, process, and distribute the majority of the nation's energy, and participate in the API Energy Excellence@ program, which is accelerating environmental and safety progress by fostering new technologies and transparent reporting. API was formed in 1919 as a standards-setting organization and has developed more than 700 standards to enhance operational and environmental safety, efficiency, and sustainability. Through the API Climate Action Framework and related initiatives such as The Environmental Partnership, significant efforts are being conducted by the oil and natural gas industry to balance the increasing demand for affordable and reliable energy products with environmental performance and stewardship.

country, with more than 530,000 miles of transmission pipelines—the majority underground—safely delivering crude oil, petroleum products, natural gas, natural gas liquids, low carbon energy products and ethanol to consumers. With abundant energy resources, America continues to chart a reliable path that directly addresses today's domestic and global energy challenges by strengthening our country's energy security while helping allies around the world.

These US energy resources have spurred community growth and security, and we have developed them while protecting the environment. The pipeline industry is developing new technologies and innovations every day to safely deliver affordable, reliable energy to consumers while reducing greenhouse gas (GHG) emissions. As our industry continues to advance a lower carbon future, it is imperative that the regulatory environment and PHMSA are responsive to both current and potential future safety challenges faced by operators.

As Congress considers the reauthorization of PHMSA and safety programs over the coming year, we urge policymakers to enact policies that capitalize on the power of American oil, natural gas, and liquified natural gas (LNG) operators by maximizing our investment in state-of-the-art technology and sustainable operations while recognizing the important role our communities play in advancing safety.

Energy Security

American energy leadership is vital, perhaps more than ever, for American prosperity and security in uncertain times, and for the world. A global energy crisis— driven by surging post-pandemic demand outstripping supply and exacerbated by Russia's invasion of Ukraine—has shown that the world needs more natural gas and oil, not less. America can and should lead the world out of this crisis. The U.S. is the world's leading producer of natural gas and oil and, with the right regulations and policies, must champion a reliable path forward that directly addresses today's energy challenges.

Europe's energy struggles show what can happen when countries lean too much, too soon on still-developing energy sources and energy from unstable regions while turning away from reliable natural gas and oil resources. Natural gas and oil are projected to provide nearly 50% of the world's energy in

2050.[74] Abundant American natural gas and oil should be prioritized as a long-term strategic asset and a foundation for economic growth and strengthened energy security, today and tomorrow. Our energy resources also will provide opportunity, for decades to come, for other energy sources to mature and take on more significant roles.

Although oil and natural gas companies increased production this year to meet rising energy demand,[75] more is required as populations and energy needs grow. America lacks sufficient infrastructure for the future—including new natural gas and oil pipelines as well as investment in maintaining existing infrastructure—to move energy from production areas to refineries and processing facilities and then to places where it is needed by American families and businesses. Our energy infrastructure is a critical component of the oil and natural gas supply chain, consisting of terminals, underground and above ground storage facilities, pipelines, railcars, trucks, ships, and marine vessels. Ensuring we have a robust energy infrastructure system that keeps pace with growing production and demand is essential to helping provide American families and businesses with reliable access to affordable energy. Additionally, export infrastructure is needed so that America can aid allies abroad and project leadership in global markets, while also supporting domestic production and economic benefits.

We need not return to growing dependence on foreign suppliers. Quite the opposite: American energy must lead in this space, through more domestic production and modernized infrastructure. If America doesn't lead, others will. We have the resources, yet sound energy infrastructure policymaking is critical to ensure America and other nations alike can have access to reliable US energy sources.

Commitment to Pipeline Safety

Industry's commitment to safe operations is evident by the strong safety record of the pipeline system that delivers oil, natural gas, and petroleum products. Protecting the public and the environment is the top priority for pipeline operators and a central component to pipeline design, construction, and maintenance. Ultimately, the development of a comprehensive pipeline safety system is the product of a shared commitment from key entities in the

[74] https://www.eia.gov/outlooks/ieo/consumption/sub-topic-01.php.

[75] https://www.api.org/-/media/Files/misc/Rystad-Energy-APIs-10-in-2022-Policy-Plan-Quantification-of-Policy-Impacts.

stakeholder community. The first element involves the federal and state governments, which provide the safety regulations for the industry. Next is the contribution of the industry trade associations that, with the help of other stakeholders including the public, help to develop industry standards, guidance, recommendations, and leading practices. The third key entity is individual companies who make the commitment to develop and implement effective safety programs. While each of these functions are critically important to advancing safety in the pipeline industry, the true effectiveness of the pipeline safety program exists because these functions complement one another through the coordination and collaboration of all three of these entities.

API, our allied oil and natural gas trades, and our members are fully committed to maintaining the highest standards and establishing a strong foundation with the public by continually endeavoring for improvement through enhanced safety operations. And while greater than 99 percent of oil, natural gas and their products reach their destination without incident, pipeline companies are striving to continuously improve performance and address the remaining fraction of a percent to reach our shared industry-wide goal of zero incidents. The industry's ability to continually advance the safety of oil and natural gas pipeline operations is based on three critical elements: (1) people, (2) technology, and (3) safety culture. Each of these is intertwined with the others to create a comprehensive and cohesive safety program. Education and training are constantly provided to industry employees to ensure they can operate using the latest and most advanced technologies. Similarly, employees are committed to developing a culture of safety that is continually assessed and improved. This three-pronged approach is designed first and foremost to prevent an incident from ever happening, but also ensures that the industry is prepared for any incident and can effectively respond in the rare instance that an incident occurs.

Effective Regulatory Process

API and its members appreciate the emphasis PHMSA has placed on addressing congressional mandates from prior pipeline safety reauthorizations. However, we ask PHMSA to continue to advance important industry rulemakings mandated by Congress in the PIPES Act of 2020, specifically leak detection and repair, LNG facility safety, class location changes, and pipeline operating status for "idle" pipe. In doing so we ask

PHMSA to maintain sight of the importance of a holistic, performance-based regulatory approach that maximizes the industry's ability to use the latest advances in new technologies and leading engineering practices to manage pipeline safety risk. Furthermore, as part of that process, we support the collaborative approach to reviewing regulations carried out by PHMSA's Technical Advisory Committees (Advisory Committee). The Advisory Committee process is a transparent and balanced forum that has demonstrated the ability to build consensus around complex regulatory issues. To that end, we have concern with the limited number of Advisory Committee meetings held over the last several years and the number of Advisory Committee recommendations that have not been incorporated into final regulations despite consensus within the groups. As such, we encourage Congress to consider the inclusion of some accountability measures within reauthorization, including a set number of meetings required each year and reporting by the Secretary on those Advisory Committee recommendations that are not being considered by PHMSA in rulemaking.

Additionally, performing a reasoned cost-benefit analysis before making significant regulatory changes must continue to be a part of the regulatory process. PHMSA's cost-benefit analyses provide valuable input to the public comment and advisory committee review processes. Since there are usually multiple practical alternatives to achieve any particular pipeline safety objective, the cost-benefit analysis helps PHMSA, and industry and public stakeholders, compare and contrast the alternatives and identify the best option.

A statutory requirement to consider costs and benefits in health, safety, and environmental regulations is not unique to PHMSA. Congress has similarly required the Occupational Safety and Health Administration (OSHA), the Mine Safety and Health Administration (MSHA), and the Environmental Protection Agency (EPA) to analyze costs and benefits during rulemaking. An excellent example of the important role cost-benefit plays in the regulatory process is PHMSA's consideration of class location changes for natural gas transmission pipelines through rulemaking. With today's processes and technologies, pipeline safety can be managed effectively and at an equivalent level of safety through data-driven inspection and maintenance, instead of costly, unnecessary, and arbitrary pipe replacements required by the current class location change regulations.

Well Trained and Efficient Regulator

As PHMSA and the energy industry together continue to drive toward our shared goal of zero pipeline incidents, a modernized regulator with the necessary tools, well-trained staff, streamlined programs and data transparency can bring needed certainty and consistency into the regulatory and oversight process. The oil and natural gas industry believes that PHMSA pipeline inspectors must be well trained and qualified to effectively fulfill the obligations placed upon them and the agency. However, pipeline inspectors frequently come into PHMSA with limited pipeline safety experience, and those that already have or gain experience often depart the agency to pursue more lucrative opportunities. As such, similar to other agency hiring authority for specialty positions, the ability to compensate pipeline inspectors at market rates through PHMSA's use of Schedule A employees with streamlined hiring and flexible pay levels would enhance PHMSA's ability to attract and retain expert pipeline inspectors.

Lastly, multiple repetitive and redundant inspections of operator procedures and programs are conducted by PHMSA regional offices, state pipeline regulatory agencies, and local authorities (e.g., state utility boards, public utility commissions) that evaluate a consistent set of company procedures regardless of jurisdiction. As an industry, we have concerns with how this multitude of agencies collaborates and communicates and the impact it can have on pipeline operators who are focusing resources on improving safety programs. As such, Congress should consider an independent review and evaluation of PHMSA and State inspection programs' interaction to identify opportunities for improved collaboration and to reduce redundant inspection activities.

The Importance of API Pipeline Standards

Safety is a core value of the oil and natural gas industry. Pipeline operating companies are committed to enhancing the safety of our workers and protecting the communities and environments where we are present. At API, we establish and disseminate standards and best practices across the industry to ensure the highest level of safety and achieve our collective goal of operating with zero incidents. Since 1924, API has been the leader in developing voluntary, consensus-based, and internationally recognized industry standards that promote safety and reliability. Our standards program is accredited by the American National Standards Institute (ANSI), the same organization that accredits similar programs at several national laboratories. In creating these industry consensus standards and recommended practices (RPs), API partners with technical experts from government, academia, and industry. This work supports the fulfillment of the National Technology Transfer and Advancement Act (NTTAA), which mandates that federal agencies use technical standards developed and adopted by voluntary consensus standards bodies, as opposed to using government-unique standards. Currently, API has more than 800 standards that are used globally by oil and natural gas operators, as well as referenced by the international regulatory community. They are key to the global energy sector, streamlining regulation across borders, improving safety, and minimizing economic inefficiencies. Here in the US, these standards are referenced more than 650 times in federal regulations, covering multiple government agencies, including PHMSA. Additionally, API's standards are the most widely cited petroleum industry standards by state regulators, with 240 API standards cited over 4,130 times in state-based regulations. Finally, API's standards are also the most widely cited standards by international regulators in the 14 major producing regions.[76]

Despite the current lack of certainty in the regulatory process, the industry continues to proactively advance safety through the development of standards and best practices. API continues to develop and revise critical standards and recommended practices for prevention, mitigation, and response activities to address pipeline safety. Specifically, API has developed a number of standards in close coordination with subject matter experts from government, academia and industry. API RP 1173, Pipeline Safety Management Systems (Pipeline SMS), provides the framework for managing complex operations with safety as the top priority. It provides established guidelines for pipeline operating companies and contractors to manage risk, promote best practices, and build a strong safety culture throughout an organization. Safety culture must be organically strengthened from within an organization, which is why a voluntary regime is so important for the industry's implementation of SMS.

Recognizing the role of stakeholder input and public engagement as critical elements in addressing pipeline safety challenges, industry has partnered with government representatives and public advocates to develop API RP 1185, Pipeline Public Engagement. The document follows the same "Plan-Do-Check-Act" cycle of continuous improvement as in API RP 1173 to

[76] OGP Report No. 426, Regulators' Use of Standards, March 2010.

regularly assess and improve public engagement in the communities where pipelines operate, including environmental justice populations. As U.S. production continues to grow and pipeline capacity increases to keep pace, operators are motivated to develop a management system that ensures new pipelines are built to the appropriate specifications, keeping safety as the top priority. API RP 1177, Steel Pipeline Construction Quality Management Systems, outlines the steps needed for constructing safe steel pipelines, from purchasing the correct material to completing the safety and integrity inspections prior to initiating operation. Additionally, API continues to develop new standards and revise and update current standards that are focused on maintaining the integrity of pipeline infrastructure and ensuring continued safe operation of pipeline systems. Examples include RP 1176 (Assessment and Management of Cracking in Pipelines), RP 1181 (Pipeline Operating Status), RP 1183 (Assessment and Management of Dents in Pipelines), and an RP under development for geohazard integrity management (RP 1187).

While pipeline operators are taking significant steps to meet the goal of zero incidents, they must have a comprehensive mitigation strategy to reduce the impact should a release occur. Developed with industry, regulator, and broader stakeholder input, API RP 1175, Pipeline Leak Detection—Program Management, outlines how to use multiple leak detection tools—such as aerial overflights, ground patrols, and computational pipeline monitoring—to create a robust and holistic program to identify a leak as soon as it occurs. In addition, the RP encourages senior leaders within pipeline operating companies to instill a leak detection culture that promotes safety and ensures employees are properly trained to aid in mitigating incidents. Pipeline companies have robust operator qualification (OQ) programs that ensure their personnel and contractor personnel are competent to perform pipeline operations and maintenance duties, and continuous training and testing to verify the skills of qualified employees is an ongoing and critical effort of operators. API has also developed RP 1161, Pipeline Operator Qualification, to give operators direction on ensuring those individuals performing important tasks that relate to pipeline operations and safety are appropriately trained and competent.

Should an incident occur, pipeline operators are ready to respond. Through coordinated emergency response programs with federal, state and local first responders and agencies, operators ensure timely, seamless, and effective responses. API RP 1174, Onshore Hazardous Liquid Pipeline Emergency Preparedness and Response, which was developed with input from operators, regulators, and first responders, seeks to improve emergency response capabilities by providing a management system and framework for operators to ensure they are prepared to respond to any event and coordinate response actions with both our government and first responder partners in an efficient manner. RPs 1161 and 1174 are just a few of the available documents developed in collaboration with federal and state regulators, academics and other interested stakeholders, which through effective implementation and training will help improve safety across the industry.

Pipeline Safety Reauthorization Priorities

As stated earlier, to improve upon our strong safety record and reach our goal of zero pipeline incidents, it is imperative that the regulatory environment and PHMSA be positioned to meet current and future safety challenges. As such, there are three priority areas where PHMSA reauthorization can support the shared objective of industry and the regulating agency in advancing pipeline safety.

1. Recognizing Role of Technology, Innovation, and Leading Practices in Reaching Zero

As an industry dependent on technology and leading engineering practices, pipeline operators want flexibility to implement and utilize the latest tools, methodologies and industry standards to appropriately manage safety risks on their facilities. However, companies are, in many instances, unable to utilize the latest technology and leading engineering practices because of the requirements associated with outdated regulations coupled with the burdensome approval process in the use of alternative safety technology. To facilitate the demonstration of an operator's effective use of state-of-the-art technology and risk-based approaches, the PIPES Act of 2020 included a provision that established a technology pilot program. Although a step in the right direction, the onerous nature of the program application requirements imposed by PHMSA proved to be an impediment to operators pursuing opportunities for testing new technology, and counter to the intent of the Congressional mandate. However, we believe the program still has merit and should be reauthorized in 2023 but more importantly modified to include more streamlined parameters that encourage, rather than discourage, an operator to participate.

For decades, a strong culture of sharing and learning has been at the center of our industry's efforts to advance pipeline safety. This includes individual company and peer to peer shares as well as annual industry wide events to openly share information and lessons learned across a broad range of topic areas, including pipeline system incidents, near misses, industry best practices, and improvements made to pipeline safety programs through practical experiences. This commitment and willingness to proactively share information on the part of our operators has been an effective tool to drive continuous improvement across the industry. That said, over time, the industry has evolved to recognize that there are even greater opportunities to learn from one another. As such, we continue to support the establishment of a more formal voluntary information sharing framework with the right protections, similar in nature and function to the successful Federal Aviation Administration and airline industry's voluntary information sharing framework. This multi-stakeholder center of excellence for the energy pipeline sector based on the proposed framework from the PHMSA Voluntary Information Sharing (VIS) effort from the PIPES Act of 2016, is a starting point. Congress should consider including language that encourages the use of academic institutions to house or play a role in the new Center location, in particular, institutions that have strong energy programs.

The PHMSA National Pipeline Mapping System (NPMS) represents the system of record upon which pipeline operators determine whether a pipeline segment directly intersects or "could affect" a High Consequence Area (HCA) and, as a result, which segments of a pipeline system require an Integrity Management Program (IMP). Recent experience of pipeline operators in updating HCA analysis for their systems has shown errors in the NPMS data layers for certain HCA categories, and questions have been raised about the sources of scientific studies and frequency of updating the information used to develop the data layers. Transparency and improvements in the process followed by PHMSA need to be provided, and industry review and participation in the process should be considered.

Lastly, there are more than 650 API standards referenced in Federal regulation. As these standards are improved through the American National Standards Institute (ANSI)-accredited process at a minimum of every 5 years, Federal regulations where standards were deemed appropriate for incorporation by reference, often are unable to be updated in a timely manner. As such, they do not keep pace with advances in pipeline safety technology and modern engineering practices incorporated into leading industry standards. Currently, approximately 50 percent of the instances where PHMSA cites API standards are not referencing the most recent version of those standards. For those API standards incorporated by reference that are updated, PHMSA should execute a more frequent review that can use the existing rulemaking process to incorporate the latest edition. We ask that Congress consider a requirement for PHMSA to ensure regulations are updated to reflect the latest edition of incorporated by reference standards within two years of completion.

2. Proactively Advancing a Safe and Sustainable Energy Future

Pipeline operators continue to address the challenge of climate change while providing affordable, safe and reliable energy. The oil and natural gas industry is also committed to building tools and implementing practices to enhance sustainability throughout its operations, and advancing a lower carbon future while reliably delivering energy products to meet increasing demand in the US, and low-carbon solutions are critical to achieving this goal. As the industry continues to invest in energy innovations like hydrogen (H2) and renewable natural gas (RNG) that will fuel the low-carbon future, a robust carbon dioxide (CO2) infrastructure system, including pipelines, is vital to meeting environmental commitments to reduce greenhouse gas (GHG) emissions.

As we embrace a low carbon energy future, it is imperative that pipeline safety regulations reflect current safety risks, consider existing industry standards pertaining to CO2 and hydrogen pipeline construction and operation, and incorporate the latest findings from PHMSA-sponsored and industry funded research and development for new pipeline systems and repurposing of existing energy pipeline infrastructure. Regulatory updates should consider an assessment of the adequacy of Parts of 49 CFR that specifically address CO2 pipeline safety for all phases of CO2 in pipeline transportation (i.e., supercritical, liquid, gas). As industry continues to develop new standards and Recommended Practices for CO2 pipeline safety, it will also be important for PHMSA to consider and ensure timely incorporation by reference into regulations. As a standard setting organization, API strongly supports the development of updates to CO2 pipeline safety regulations to address transporting CO2 in a gas phase based on existing standards developed through API, the American Society of Mechanical Engineers (ASME), the Association for Materials Protection and Performance (AMPP), and the International Standards Organization (ISO). The update to these regulations will require close coordination with Department of Energy (DOE), PHMSA, and Pipeline Research Council International (PRCI) to ensure incorporation of research and technology development programs that address design, construction, operations, and inspection and maintenance, and to defer development of new regulations until the research is completed and results from that work are available.

Also, in carrying out its responsibility to provide safe, reliable, and affordable energy, our industry is committed to sustainable operations and risk-based inspection programs that focus on reducing harmful emissions. Specifically, as it relates to the inspection of above ground storage tanks, operators look to the latest technologies and leading engineering practices based on industry standards to identify risks without taking tanks out of service. As a result, by decreasing the frequency of visual inspections, operators are able to reduce unnecessary releases of VOCs associated with preparing tanks for inspection. However, as previously discussed, approximately 50 percent of the API standards that are cited by PHMSA in federal pipeline safety regulations are not referencing the most recent edition of those standards which includes API Standard 653, Standards for Tank Inspection, Repair, Alteration and Reconstruction and as such do not reflect current leading industry approaches for tank inspections. This includes risk-based approaches similar to what has been accepted by other federal and state agencies, including the EPA's Spill Prevention, Control and Countermeasure (SPCC) program. Thus, operators are required to take their aboveground storage tanks out of service at predetermined intervals with no added benefit to safety, which unnecessarily increases VOC emissions. PHMSA should acknowledge the value of technology and leading engineering practices, and we are asking Congress to require PHMSA through rulemaking to help reduce environmental impacts by allowing more proactive operator use of risk-based tank inspection protocols through the incorporation of the latest edition of API Standard 653 by reference.

Lastly, pipeline safety can be managed effectively through data-driven inspection and maintenance, instead of the arbitrary pipe replacements required by the current class location change regulations. Class location regulations result in unnecessary replacement projects that can disrupt natural gas service and release up to 800 million standard cubic feet of natural gas every year, which is equivalent to the annual natural gas use of over 12,000 homes. Each year the class location changes regulations and diverts hundreds of millions of dollars towards replacing less than 75 miles of pipe. There are many more productive ways to expend these substantial resources. For example, for the same cost of replacing 75 miles of pipe, we could instead assess 25,000 miles of pipelines with in-line inspection tools that would reduce the risk of potential incidents that could impact people or the environment. As such, as mandated by Congress in the 2020 PIPES Act, PHMSA needs to publish a final rule on class location changes.

3. Embracing and Recognizing Communities as Partners in Safety

As the industry expands oil and natural gas infrastructure, it is incumbent on pipeline operators to establish local relationships, address performance issues and engage community stakeholders. The industry's commitment to being a good neighbor throughout the lifecycle of a pipeline requires ongoing dialogue with local communities and other key stakeholders. API continues to spearhead efforts on public and community engagement initiatives to reinforce industry's social license to operate. The center of API's stakeholder and public engagement strategic goal continues to be the development of Recommended Practice (RP) 1185, Pipeline Public Engagement, which addresses critical topics such as environmental justice, engagement with landowners and tenants, and other issues concerning effective community engagement throughout the entire lifecycle of a pipeline.

Additionally, API recently published the third edition of RP 1162, Public Awareness Programs for Pipeline Operators and is in the process of developing an implementation guidance website. The 3rd edition includes significant updates that reflect more forward leaning industry approaches to engaging with and raising public and stakeholder awareness of pipeline operations and important safety programs along pipeline right of ways that protect the public and environment. Specifically, this provision expands the current requirements of pipeline operator public awareness programs, incorporating the latest mechanism for sharing information, expanding risk communications, and establishing processes for evaluating the effectiveness of programs. API strongly encourages PHMSA to initiate a rulemaking to consider the inclusion of the latest edition of RP 1162 (3rd edition).

Lastly, API remains committed to building tools and platforms to help the industry enhance and expand a culture of safety and sustainability across all operations. As such, API published conservation guidelines for right of way (ROW) ecological and habitat management and is working with member companies in partnership with NGOs, such as Pheasants Forever, in implementing conservation projects throughout the United States. Built upon adaptive management principles, the API Conservation Program takes an integrated and systematic approach in sustaining ROW land management that is value-driven and enhances safety, community benefits, operational efficiencies, and a healthier ecosystem while maintaining state and federal regulatory compliance and considering associated risk and cost. As pipeline operators continue to advance conservation programs, it is important that the regulations recognize these alternative approaches and allow pipeline operators to maintain right of ways using non-traditional methods including conservation, habitat management and other related programs while ensuring pipelines can be effectively monitored through required surveys. Conservation and habitat management programs may require alternatives to existing requirements for line markers and leak detection methodologies, etc. As such, API would like changes to statute and ultimately regulations that recognize the importance of approaching right of way maintenance in non-traditional ways, while still meeting regulatory requirements for safety.

Conclusion

Safety of the public and the environment is our industry's top priority, and collaboration with PHMSA only strengthens our ability to transport our products across America with the fewest possible number of incidents. We are committed to promoting safety in all our operations, helping to ensure that American families and businesses can efficiently access affordable and reliable energy. Thank you for the opportunity to provide comments in advance of the hearing and deliberations on pipeline safety reauthorization.

Press Release of March 8, 2023, from the Distribution Contractors Association and the United Association of Union Plumbers and Pipefitters, Submitted for the Record by Hon. Troy E. Nehls

FOR IMMEDIATE RELEASE
March 8, 2023
Contact:
Eben M. Wyman (703) 740–6126
eben@wymanassociates.net

Construction Entities Call for Punishment for Pipeline Attacks, Improved Mapping in Pipeline Safety Hearing

Richardson, TX—The Distribution Contractors Association (DCA) and the United Association of Union Plumbers and Pipefitters (UA) urge Congress to take action to address the enduring problem of physical attacks on critical energy infrastructure. In addition, Congress should include provisions to encourage the use of geographic information systems (GIS) mapping as part of these damage prevention programs. On Wednesday, the House Transportation and Infrastructure Committee's Subcommittee on Railroads, Pipelines, and Hazardous Materials held a hearing on "Pipeline Safety: Reviewing Implementation of the PIPES Act of 2020 and Examining Future Safety Needs."

The hearing was held to evaluate efforts to meet the mandates included in the Protecting Our Infrastructure of Pipelines and Enhancing Safety (PIPES) Act of 2020 at a time when the Committee intends to reauthorize the federal pipeline safety program this year. These organizations are advocating on a range of issues pending in this year's pipeline safety discussion, including suggested actions to curb physical attacks on energy infrastructure and to encourage the use of GIS mapping.

Protection of Critical Infrastructure from Physical Attacks

"While criminal attacks on pipeline infrastructure is not headline news anymore, sabotage on natural gas pipelines is an enduring problem and needs to be addressed," said Rob Darden, Executive Vice President of DCA. "While DCA supports the right for peaceful activism, including peaceful protests, we strongly support legislative language that would hold those who engage in criminal activities during protests more accountable."

"In addition, pipeline facilities under construction should be included within the scope of this provision. While interfering or tampering with the operation of a pipeline would clearly compromise pipeline safety, vandalism and destruction of nearby equipment used to build a pipeline can be just as dangerous," says Darden. "For example, setting construction equipment on fire near a natural gas pipeline can be as dangerous as turning a valve."

"United Association members are the best trained and most highly skilled craftspeople in the industry—and we work tirelessly to provide reliable and affordable energy to millions of Americans. Sadly, physical attacks on pipeline infrastructure itself and even our equipment and jobsites compromise our shared goals and put the safety of our members at risk," said Mark McManus, General President of the United Association. "Strong bipartisan action to hold perpetrators accountable, and to curb future attacks, is essential to protecting our members on the jobsite and ensuring we can meet our shared energy goals."

GIS Mapping

The PIPES Act of 2020 include provisions to require operators of gas distribution pipelines to "identify and manage traceable, reliable, and complete records, including maps and other drawings," and language that will ensure that this documentation is "accessible to all personnel responsible for performing or overseeing relevant construction or engineering work." DCA and the UA believe increased use of GIS is the most effective method of mapping underground facilities, including natural gas pipelines.

"Accurate mapping of underground facilities is a fundamental part of accurate and timely locating,'' Darden said. ''Ensuring the use of GIS should be considered a priority as pipeline safety legislation is developed and considered.''

GIS connects data to a map, integrating location data with a range of limiting information regarding the subsurface facilities in that area, and it allows for layering of data tied to geographic points. Rather than restricting the user to limited features on a static map, GIS mapping allows for viewing customizable combinations of data layers in a single dynamic tool.

###

The Distribution Contractors Association represents contractors, suppliers and manufacturers who provide distribution construction services including installation, replacement and rehabilitation of gas pipelines and fiber optic, cable and duct systems in communities across the country.

Founded in 1889, the United Association of Journeymen and Apprentices of the Plumbing and Pipe Fitting Industry of the United States and Canada, AFL–CIO, CLC (UA) proudly represents more than 362,000 highly skilled journeymen and apprentice plumbers, pipefitters, pipeliners, sprinkler fitters, welders, and HVACR service techs in the US and Canada. The UA is an affiliate of the AFL–CIO, North America's Building Trades Unions (NABTU), and the Canadian Labour Congress (CLC).

Mr. NEHLS. In closing, I want to thank our witnesses for coming today. I appreciate you being here.

I will now yield back and recognize Ranking Member Payne for 5 minutes for an opening statement.

Opening Statement of Hon. Donald M. Payne, Jr., of New Jersey, Ranking Member, Subcommittee on Railroads, Pipelines, and Hazardous Materials

Mr. PAYNE. Thank you, Chairman Nehls. It is an honor and a privilege to serve on this committee with you, and I am looking forward to your leadership and the bipartisan manner which you have already demonstrated. It is greatly appreciated.

I would like to also thank Chairman Graves for his help, and also thank Ranking Member Larsen for being here, and all the witnesses that are with us today. Thank you.

Before I get to the subject of today's hearing, I want to note that this is the first hearing of the Subcommittee on Railroads, Pipelines, and Hazardous Materials this Congress, and the first following Norfolk Southern derailment in East Palestine, Ohio. This has been followed by additional Norfolk Southern derailments in southeast Michigan and Springfield, Ohio. Just this week, a Norfolk Southern employee lost his life in a collision in Cleveland.

Chairman Nehls, thank you for visiting East Palestine immediately following the derailment. I look forward to working with you this Congress to tackle freight and rail safety.

I am pleased that the Senate has put forth a bipartisan freight rail safety bill, and I am hopeful that we can do the same on this side.

The safe operation of pipelines is vital to protect both our energy needs and the communities these pipelines run through.

These are not just abstract concerns. Just last December, a pipeline operated by TC Energy ruptured in northeast Kansas, spilling over 14,000 barrels of crude oil. If you have a hard time visualizing 14,000 barrels, it's the same volume as 20 rail tank cars.

In Linden, New Jersey, in my district, a pipeline operated by Buckeye Partners ruptured and spilled 350 barrels adjacent to a fragile wetland in March 2021.

We must learn lessons from these incidents as well as the Denbury incident in Satartia, Mississippi, the Beta Offshore incident in San Pedro Bay, California, and the Freeport LNG incident in Quintana Island.

PHMSA reports that in 2022, 631 pipeline incidents occurred, from which 10 individuals died, 24 were injured, and property damages totaled more than $680 million. Of these incidents, 17 were serious and 267 were significant.

I look forward to hearing from our witnesses today about their commitment to safety, if PHMSA has the resources necessary to hold bad actors accountable, and how can we make sure that what goes into pipelines stays in the pipes.

I am also interested to hear about PHMSA's progress on carbon dioxide pipeline rulemaking.

The safe transmission of carbon dioxide to sequester locations is vital to meeting our carbon reduction goals, and I want to make sure this can be implemented without delay.

Thank you, Mr. Chairman, for the time, and I yield back. [Mr. Payne's prepared statement follows:]

Prepared Statement of Hon. Donald M. Payne, Jr., a Representative in Congress from the State of New Jersey, and Ranking Member, Subcommittee on Railroads, Pipelines, and Hazardous Materials

Thank you, Chairman Nehls, Chairman Graves, Ranking Member Larsen, and all our witnesses for being with us today.

Before I get to the subject of today's hearing, I want to note that this is the first hearing of the Railroad Subcommittee this Congress and the first following the Norfolk Southern derailment in East Palestine, Ohio.

This has been followed by additional Norfolk Southern derailments in Southeast Michigan and Springfield, Ohio Just this week a Norfolk Southern employee lost his life in a collision in Cleveland.

Chairman Nehls, thank you for visiting East Palestine immediately following the derailment.

I look forward to working with you this Congress to tackle freight rail safety.

I'm pleased that the Senate has put forth a bipartisan freight rail safety bill and I'm hopeful that we can do the same over here.

The safe operation of pipelines is vital to protecting both our energy needs and the communities these pipelines run through.

These are not just abstract concerns.

Just last December, a pipeline operated by TC Energy ruptured in northeast Kansas, spilling over 14,000 barrels of crude oil.

If you have a hard time visualizing 14,000 barrels, it's the same volume as 20 rail tank cars.

In Linden, New Jersey, in my district, a pipeline operated by Buckeye Partners ruptured and spilled 350 barrels adjacent to a fragile wetland in March 2021.

We must learn lessons from these incidents as well as the Denbury incident in Satartia, Mississippi, the Beta Offshore incident in San Pedro Bay, California, and the Freeport LNG incident in Quintana Island.

PHMSA reports that in 2022, 631 pipeline incidents occurred, from which 10 individuals died, 24 were injured, and property damages totaled more than 680 million dollars.

Of these incidents, 17 were serious and 267 were significant.

I look forward to hearing from our witnesses today about their commitment to safety, if PHMSA has the resources necessary to hold bad actors accountable, and how we can make sure that what goes into pipelines stays in the pipes.

I'm also interested to hear about PHMSA's progress on carbon dioxide pipeline rulemaking.

The safe transmission of carbon dioxide to sequester locations is vital to meeting our carbon reduction goals, and I want to make sure this can be implemented without delay.

Thank you again Mr. Chairman, and I yield back my time.

Mr. NEHLS. Thank you, Mr. Payne. I appreciate that. I now recognize the ranking member of the full committee, Mr. Larsen, for 5 minutes for an opening statement.

Opening Statement of Hon. Rick Larsen of Washington, Ranking Member, Committee on Transportation and Infrastructure

Mr. LARSEN OF WASHINGTON. Thank you. Thank you, Chair Nehls and Ranking Member Payne, for calling this hearing on pipeline safety.

On June 10, 1999, a pipeline explosion in Bellingham, Washington, claimed the lives of two 10-year-old boys and a young man of 18 years. The explosion released 237,000 gallons of gasoline into Whatcom Creek that flowed through Whatcom Falls Park in Bellingham, Washington. The ensuing fireball caused millions of dollars of damage, shocked the community that day, and is still a seminal moment in the history of the town of Bellingham.

It is that explosion that spurred my commitment, which has been steadfast over the last 20 years, to the highest level of pipeline safety. For my entire tenure in Congress, I have fought to reduce the risk of pipeline incidents, promote transparency of pipeline safety information for local communities, and increase accountability for pipeline operators.

Mr. Chair, I ask unanimous consent to submit a letter for the record from a parent of one of the boys killed in the explosion in my district, Bellingham City Council Member Skip Williams. This letter outlines a number of recommendations to improve the safety and oversight of our pipeline system. So, I ask unanimous consent to enter that in the record.

Mr. NEHLS. Without objection, so ordered. [The information follows:]

Letter of March 7, 2023, to Hon. Troy E. Nehls, Chairman, and Donald M. Payne, Jr., Ranking Member, Subcommittee on Railroads, Pipelines, and Hazardous Materials, and Hon. Sam Graves, Chairman, and Hon. Rick Larsen, Ranking Member, Committee on Transportation and Infrastructure, from Edwin H. "Skip" Williams, Bellingham City Council (Washington State), Submitted for the Record by Hon. Rick Larsen

MARCH 7, 2023.

The Subcommittee on Railroads, Pipelines, and Hazardous Materials of the Committee on Transportation and Infrastructure,

Representative TROY NEHLS, Subcommittee Chair,
Representative DONALD PAYNE, Subcommittee Ranking Member,
Representative SAM GRAVES, Transportation and Infrastructure Committee Chair,
Representative RICK LARSEN, Transportation and Infrastructure Committee Ranking Member,
Rayburn House Office Building, Washington, DC.

RE: Reviewing implementation of the PIPES Act of 2020 and Examining Future Pipeline Safety Needs

Dear Subcommittee Members,

I am Edwin H "Skip" Williams, a 30-year resident of the city of Bellingham, Washington, currently serving as a member of the Bellingham City Council. Thank you for allowing me to address your subcommittee hearing voicing my support for the overall need of pipeline safety and, more specifically, the elements addressed in the implementation of the PIPES Act of 2020 and your examination of future pipeline needs.

My family and I are victims of the Olympic Pipeline explosion that occurred in our city in 1999 where we lost one of our children. When the Olympic pipeline exploded there seemed to be very few regulations, industry practices, or oversight regarding pipeline safety. The elements that did exist did not seem to be enforced very consistently if at all. This event in our community highlighted the need for research, consistent industry safety standards regarding best practices, and oversight from all levels of government in order to enforce these safety standards.

I strongly believe that there needs to be open communications between pipeline operators, local communities, and federal and state governmental agencies in order to avoid pipeline accidents such as the Olympic Pipeline explosion.

As I read through the summary of the issues being considered by the committee for implementation of the PIPES Act of 2020 and the committee's examination of future safety needs, I see the following as priorities:

- Providing greater resources at the state and local level to increase the number of pipeline inspectors.
- Eliminating the backlog of rulemaking through greater congressional oversight and additional resources.

- Minimizing leaks and releases into our environment by requiring updated inspection and maintenance plans of pipeline operators.
- The study of methods to improve pipeline integrity in order to address risks posed by ageing cast iron pipes and low pressure systems.

I feel strongly that these are all significant steps forward improving the oversight protections and maintaining the integrity of our pipeline system.

I greatly appreciate the opportunity to share my perspective on pipeline safety with the subcommittee. If you would like to discuss these concerns with me further, please feel free to contact me.

Sincerely,
EDWIN H. "SKIP" WILLIAMS,
Bellingham City Council.

Per Washington state law RCW 42.56, my written communications are public records and are subject to public disclosure requirements.

Mr. LARSEN OF WASHINGTON. I share many of Council Member Williams' views, and as ranking member, I will ensure that this committee's work to reauthorize the PIPES Act of 2020 puts safety at the forefront of every policy decision.

Putting safety first means greater oversight and accountability of the activities of pipeline operators. It means greater transparency for local communities and the public. Reducing the risk of incidents means applying safety requirements, where appropriate, to existing pipeline infrastructure. PHMSA, the agency, cannot effectively doits job if the infrastructure already in the ground is off limits to safety regulation.

Improving safety means preventing incidents. PHMSA needs to have the resources and staff to inspect pipelines, conduct investigations if incidents occur, and take appropriate enforcement action. I am especially pleased that we will hear detailed recommendations on what actions Congress can take to advance safety from a fellow Washingtonian and constituent from Whatcom County on today's panel, Bill Caram, executive director of the Pipeline Safety Trust. The Trust was formed following the deadly Olympic pipeline explosion in my district. I think you will hear from Bill that the Pipeline Safety Trust is both a watchdog and also can be a partner with communities and with industry to ensure long-term safety.

I appreciate each of our witnesses, as well, being here today. I want to welcome PHMSA Deputy Administrator Tristan Brown, who visited the district to tour the Olympic pipeline site in Bellingham.

PHMSA has a new charge, as well, since the enactment of the Bipartisan Infrastructure Law, to distribute $1 billion over 5 years under the first-ever Natural Gas Distribution Infrastructure Safety and Modernization grant program. The $200 million available to municipalities and community-owned utilities in fiscal year 2022 is the first installment of this funding to repair or replace natural gas pipelines and help reduce incidents and improve safety. And this is just the beginning.

All safety responsibility, though, must not fall to PHMSA. I encourage the involvement of communities and public interest organizations in pipeline safety, and I support ongoing grants to support the 400 State-based safety inspectors of intrastate pipelines and local distribution systems.

Finally, as we convene the Subcommittee on Railroads, Pipelines, and Hazardous Materials for the first time in this Congress, I do want to call attention to the request from Members on my side of the aisle, led by Ranking Member Payne and Representative Sykes, that this committee hold a hearing on the Norfolk Southern derailment in East Palestine, Ohio. I endorse and support their efforts. Pipelines play a critical role in the Nation's infrastructure and the daily lives of Americans. And we are here today to make sure the national pipeline network safely delivers essential energy products across the country. I look forward to today's discussion. [Mr. Larsen of Washington's prepared statement follows:]

Prepared Statement of Hon. Rick Larsen, a Representative in Congress from the State of Washington, and Ranking Member, Committee on Transportation and Infrastructure

Thank you, Chairman Graves, Chairman Nehls and Ranking Member Payne for calling this hearing on pipeline safety.

On June 10, 1999, a pipeline explosion in Bellingham, WA claimed the lives of two 10-year-old boys and a young man of 18 years. The explosion also released 237,000 gallons of gasoline into Whatcom Creek that flowed through

Whatcom Falls Park in Bellingham, Washington. The ensuing fireball caused millions of dollars of damage. It shocked the community that day and is still a seminal moment in the history of the town of Bellingham.

This explosion spurred my commitment, which has been steadfast for over 20 years, to the highest level of pipeline safety. For my entire tenure in Congress, I have fought to reduce the risk of pipeline incidents, promote transparency of pipeline safety information for local communities and increase accountability for pipeline operators.

I ask unanimous consent to submit a letter for the record from a parent of one of the boys killed in the explosion in my district, Bellingham City Council Member Skip Williams. This letter outlines a number of recommendations to improve the safety and oversight of our pipeline system.

I share many of Councilmember Williams' views and as Ranking Member, I will ensure that this Committee's work to reauthorize the PIPES Act of 2020 puts safety at the forefront of every policy decision.

Putting safety first means greater oversight and accountability of the activities of pipeline operators. It also means greater transparency for local communities and the public.

Reducing the risk of incidents means applying safety requirements, where appropriate, to existing pipeline infrastructure. PHMSA cannot effectively do its job if infrastructure already in the ground is off limits to safety regulation.

Improving safety means preventing incidents. PHMSA needs to have the resources and staff to inspect pipelines, conduct investigations if incidents occur, and take appropriate enforcement action.

I am especially pleased that we will hear detailed recommendations on what actions Congress can take to advance safety from a fellow Washingtonian—and constituent from Whatcom County—on today's panel: Bill Caram, Executive Director of the Pipeline Safety Trust. The Trust was formed following the deadly Olympic Pipeline explosion in my district. I think you will hear from Bill that the Pipeline Safety Trust is both a watch dog and can also be a partner with communities and with industry to ensure long-term safety.

I appreciate each of our witnesses being here today. I want to welcome PHMSA Deputy Administrator Tristan Brown, who visited Washington's Second District to tour the Olympic pipeline site in Bellingham.

PHMSA has a new charge, since the enactment of the Bipartisan Infrastructure Law (BIL), to distribute $1 billion over five years under the first ever Natural Gas Distribution Infrastructure Safety and Modernization grant program. The $200 million available to municipalities and community-owned utilities in FY 2022 is the first installment of this funding to repair or replace natural gas pipelines and help reduce incidents and improve safety. This is just the beginning.

All safety responsibility must not fall to PHMSA. I encourage the involvement of communities and public interest organizations in pipeline safety. I support ongoing grants to support the 400 state-based safety inspectors of intrastate pipelines and local distribution systems.

Finally, as we convene the Subcommittee on Railroads, Pipelines, and Hazardous Materials for the first time in this Congress, I call to attention the request from Members on my side of the aisle, led by Ranking Member Payne and Rep. Sykes, that this Committee hold a hearing on the Norfolk Southern derailment in East Palestine, Ohio. I endorse and support their efforts.

Pipelines play a critical role in the nation's infrastructure and the daily lives of Americans. We are here today to make sure the national pipeline network safely delivers essential energy products across the country. I look forward to today's discussion.

Mr. LARSEN OF WASHINGTON. With that, I yield back. Thank you, Mr. Chair.

Mr. NEHLS. Mr. Larsen yields. Again, I would like to welcome our witnesses, and thank you for being here today.

I ask unanimous consent that the witnesses' full statements be included in the record.

Without objection, so ordered.

As your written testimony has been made part of the record, the subcommittee asks that you limit your oral remarks to 5 minutes. With that, Deputy Administrator Brown, you are recognized for 5 minutes for your testimony.

Testimony of Tristan Brown, Deputy Administrator, Pipeline and Hazardous Materials Safety Administration (PHMSA); Andrew J. Black, President and Chief Executive Officer, Liquid Energy Pipeline Association (LEPA); Kenneth W. Grubb, Chief Operating Officer, Natural Gas Pipelines, Kinder Morgan, Inc., on Behalf of the Interstate Natural Gas Association of America (INGAA); and Bill Caram, Executive Director, Pipeline Safety Trust

Testimony of Tristan Brown, Deputy Administrator, Pipeline and Hazardous Materials Safety Administration (Phmsa)

Mr. BROWN. Thank you, Chairman Nehls. Thank you, Ranking Member Payne. Thank you, Ranking Member Larsen. And to the rest of the committee, welcome to the new Members, and welcome back to the returning Members.

At the Department of Transportation and within PHMSA, safety is the top priority of Secretary Buttigieg, Deputy Secretary Trottenberg, myself, and all of the employees at PHMSA. We all appreciate this subcommittee's interest and support for strengthening pipeline safety across our country.

PHMSA is responsible for overseeing the safe transport of hazardous materials through pipelines and also via other modes of transportation, including planes, trains, trucks, automobiles, and vessels. On that note, although I am here with a panel of pipeline experts to discuss PHMSA's work in advancing pipeline safety, I want to take this chance to acknowledge that, as an agency, as a Department, and across the Federal Government, we remain focused on holding Norfolk Southern accountable for the terrible tragedy in East Palestine, Ohio. And we will continue to provide support to that community and to the National Transportation Safety Board's investigation.

Over the years, despite oftentimes fierce opposition from the industries we regulate, PHMSA has worked to strengthen safety measures for transporting hazardous materials, and we look forward to working with Members on both sides of the aisle to continue those efforts by statute, by regulation, or any other means of achieving results for the people of East Palestine and many other communities that have suffered similar incidents in the past.

PHMSA's oversight of hazardous materials via other modes includes nearly 1 in 10 goods that are transported in the United States, everything from nuclear waste to lithium-ion batteries to explosives used in excavation mining and energy production. In the United States, nearly two-thirds of the energy we consume is transported via a pipeline that we regulate.

Over the past few decades, growth in energy production in the United States has increased to record levels over the last few years. Concurrently, U.S. transportation of these products has increased, and exports of energy have also reached record levels. This has placed new and heightened demands on our pipeline and refined product storage infrastructure, as well as export facilities such as LNG terminals, which PHMSA also regulates.

Under Secretary Buttigieg's leadership, PHMSA has been focused on executing bipartisan congressional mandates in the PIPES Act of 2020, historic infrastructure investments that came from the Bipartisan Infrastructure Law of 2021 that Ranking Member Larsen mentioned, strengthening our safety mission, and ensuring the U.S. maintains the most efficient and competitive transportation system in the world.

The bipartisan PIPES Act of 2020 significantly strengthened PHMSA's jurisdiction related to the minimization of methane emissions across all of our regulated entities in an effort to improve public safety and protect our environment. Our efforts on this front include completing three major legacy pipeline safety rulemakings, each of which were more than a decade in the making, including new regulations on 400,000 miles of gas gathering pipelines, pipeline rupture detection and rupture mitigation valve installation, and Coastal Ecological Unusually Sensitive Areas, including the Great Lakes.

Our regulatory agenda over the past 2 years has been exceptionally full. PHMSA has been hard at work closing out years' long— and in some cases, decades' long—efforts on final rules, as well as initiating important new rulemakings from the 2020 PIPES Act.

However, with all of the good work that is being done to advance pipeline safety by the promulgation of new rules, PHMSA has faced a new normal in terms of increased challenges to its rulemakings. This is resulting in longer development timelines and diversion of personnel resources to respond to legal challenges. That time could otherwise be utilized to advance the myriad congressional directives and regulatory priorities of the agency and stakeholders.

Regardless of these challenges, PHMSA won't be swayed from its mission. PHMSA has initiated several priority rulemakings that look to address important PIPES Act mandates and emerging safety issues, including leak detection and leak repair, the safety of gas distribution pipelines, updated LNG facilities regulations, and an overhaul of the safety requirements for pipelines transporting carbon dioxide.

As previously noted, PHMSA's oversight responsibilities continue to grow, both in terms of the types of facilities we regulate, as well as the number of facilities we regulate. And we have had to continuously operate relatively leaner, as compared to our expanded universe of regulated facilities. To meet congressional directives to improve efforts to attract and retain pipeline engineers and inspectors, PHMSA has undertaken new recruitment and retention efforts, has kept up with the hiring mandates included in the PIPES Act, both for inspectors as well as for regulatory personnel, that have helped lead the agency to some of its most important, productive years ever, in terms of both finalizing rulemakings, as well as enforcement actions, and has reduced the trend in hazardous materials pipeline incidents.

In closing, thank you again for the opportunity to engage with you on the critical issues facing PHMSA and, in turn, facing a major component of the largest, most sophisticated energy transportation system in the world. We look forward to continuing to work with you to improve pipeline and hazardous materials safety, and to reduce associated environmental impacts. I look forward to your questions.

[Mr. Brown's prepared statement follows:]

Prepared Statement of Tristan Brown, Deputy Administrator, Pipeline and Hazardous Materials Safety Administration (PHMSA)

Introduction

Chairman Nehls, Ranking Member Payne, Jr., and members of the Subcommittee, thank you for inviting me to testify today on the U.S. Department of Transportation's (DOT) Pipeline and Hazardous Materials Safety Administration's (PHMSA) pipeline safety program. I appreciate this subcommittee's support for strengthening pipeline safety across our country. I am here on a panel of pipeline safety experts to discuss PHMSA's work advancing pipeline safety, but I want to acknowledge that as an agency, as a department, and across the Federal Government, we also remain focused on holding Norfolk Southern accountable for the terrible tragedy in East Palestine, OH—and will continue to provide support to that community and to the National Transportation Safety Board, the independent lead investigator. Over the years, despite often fierce opposition from the industries we regulate, PHMSA has worked to strengthen safety measures for transporting hazardous materials via rail and other modes of transportation, We look forward to working with members on both sides of the aisle to continue those efforts by statute, by regulation, or any other means of achieving results for the people of East Palestine and many other communities that have suffered similar incidents in the past.

At DOT, and within PHMSA, safety is the top priority for Secretary Buttigieg, Deputy Secretary Trottenberg, me, and all of the employees at PHMSA. Specifically, PHMSA is responsible for overseeing the safe transport of hazardous materials— through pipelines and also via other modes of transportation—including planes, trains, trucks, and vessels. PHMSA oversees the safe design, operation, and maintenance of the Nation's nearly 3.4 million miles of oil, gas, and other hazardous materials pipeline facilities for hydrogen, carbon dioxide, and other emerging fuels. Additionally, PHMSA's oversight of hazardous materials via other modes includes nearly 1 in 10 goods that are transported in the U.S., everything from nuclear waste to lithium-ion batteries, to explosives used in excavation, mining, and energy production. PHMSA also chairs the International Civil Aviation Organization's Dangerous Goods Panel, the international standards making body that sets the global framework for the safe and efficient transport of these materials across borders and around the world.

Under Secretary Buttigieg's leadership, PHMSA has been focused on executing bipartisan congressional mandates in the Protecting our Infrastructure of Pipelines and Enhancing Safety Act of 2020 (PIPES Act of 2020), historic infrastructure investments from the Infrastructure Investment and Jobs Act of 2021, strengthening our safety mission, and ensuring that the U.S. has the safest, most efficient and competitive transportation system in the world.

From the standpoint of the volume of work before us as an agency, the challenges in carrying out our safety mission have never been greater. We oversee an aging infrastructure that requires robust maintenance, and, when needed, replacement. Most of the cross-country pipeline infrastructure was built shortly after World War II—meaning many pipelines are over 80 years old—and there are even a few gas distribution segments that were installed during the Civil War era, more than 150 years ago.

PHMSA has been integral to the whole-of-government approach to mitigating unnecessary greenhouse gas emissions—an essential component of operating the safest, most efficient and economically competitive transportation and energy system of the 21st century.

Nearly two-thirds of the energy we consume in the U.S. is transported via pipeline. Over the past few decades, growth in energy production in the United States has increased to record levels. Concurrently, U.S. transportation of these products has necessarily increased, and exports of energy have—according to the Energy Information Administration—also reached record levels. This has placed new and heightened demands on our pipeline and refined products storage infrastructure, as well as export facilities, such as liquefied natural gas (LNG) terminals, which PHMSA also regulates.

The bipartisan PIPES Act of 2020 significantly strengthened PHMSA's jurisdiction related to the minimization of methane emissions across all of our regulated entities in an effort to improve public safety and protect our environment. Our efforts on this front include completing three major legacy pipeline safety rulemakings, each of which was more than a decade in the making, including new regulations on 400,000 miles of "gas gathering" pipelines—significantly increased by the fracking boom that began in the 2000s but remained unregulated at the Federal level until this past year. Since the enactment of the PIPES Act of 2020, Congress has added new incentives for infrastructure aimed at decarbonizing energy and industrial sectors. Both the Infrastructure Investment and Jobs Act and the Inflation Reduction Act include significant incentives for the build-out of the hydrogen and carbon capture utilization and storage (CCUS) sectors. To address new risks—both safety and environmental—related to this infrastructure, PHMSA has focused on strengthening its regulations and improving its research in these areas. On the other side of the agency, which focuses on hazardous materials transportation via other modes of transportation, the agency has focused on improving safety in the transportation of hydrogen and other cryogenic materials—via truck, train, plane, and vessel—which is seeing new technology development and investments from nearly every sector of the economy.

Also, since the enactment of the PIPES Act of 2020, PHMSA has worked with Congress to advance its first-ever infrastructure grant program, via the Infrastructure Investment and Jobs Act. The Natural Gas Distribution Infrastructure Safety and Modernization (NGDISM) Grant Program provides $1 billion spread over five years to improve the safety of high-risk, leak-prone legacy natural gas distribution infrastructure with a specific emphasis on benefiting underserved rural and urban communities, among other considerations. Eligible entities are municipality- or community-owned utilities, and funds are available to these entities seeking assistance in repairing, rehabilitating, or replacing high-risk, leak-prone natural gas distribution infrastructure. Funds may also be used to acquire equipment to assist in identifying and reducing natural gas distribution pipeline incidents and fatalities. This grant funding will help communities of all sizes make their infrastructure safer, create good jobs, reduce heat-trapping methane from the atmosphere, and reduce the risks of fatality and serious injury for residents and businesses. As of this hearing, we expect to announce the first round of these grants imminently.

Rulemaking

Our regulatory agenda over the past two years has been exceptionally full. PHMSA has been hard at work closing out years-long, and in some cases decade-long, efforts on final rules as well as initiating important new rulemakings from the 2020 PIPES Act.

Some of PHMSA's recently completed final rules include the sibling rules to the Safety of Gas Transmission Pipelines final rule that we published in 2019, including improved regulation of higher-risk gas gathering pipelines; the 400,000 miles of additional pipelines previously mentioned; required reporting of safety information for all gas gathering pipelines; improved repair criteria for gas transmission pipelines; and other enhanced safety requirements for gas transmission pipelines regarding corrosion control, management of change, and inspections following extreme weather events.

Additionally, in April of 2022, PHMSA published the long-awaited final rule addressing rupture detection and rupture-mitigation valve installation for many gas transmission, hazardous liquid, and gas and hazardous liquid gathering pipelines, including carbon dioxide lines. Among other important provisions, this rule requires the installation of remote-control or automatic-shutoff valves, or equivalent technology, that can close within 30 minutes of an operator being notified of a potential rupture—saving lives and reducing methane emissions.

In response to Congress' very stringent timeline for issuing a final rule for Coastal Ecological Unusually Sensitive Areas (90 days), PHMSA issued an Interim Final Rule to include additional coastal waters, the Great Lakes, and coastal beaches within the definition of an "unusually sensitive area" for the purposes of resilience and risk reduction through hazardous liquid pipeline integrity management— strengthening protections for these treasured natural environments. PHMSA subsequently held a Liquid Pipeline Advisory Committee meeting on the rulemaking and emerged from that meeting with recommendations to help finalize this rule while addressing stakeholder concerns.

PHMSA has initiated several priority rulemakings that look to address important PIPES Act of 2020 mandates and emerging safety issues, including leak detection and leak repair, the safety of gas distribution pipelines (as directed in the Leonel Rondon Pipeline Safety Act), updated LNG facilities regulations, and an overhaul of the safety requirements for pipelines transporting carbon dioxide.

As required by Congress, PHMSA continues to update the Federal pipeline safety regulations (PSRs) to reflect new and revised voluntary consensus standards developed and adopted by standards-setting bodies (see e.g., PHMSA's periodic standards update rulemakings). We understand how important updating and aligning standards can be to ensure the PSR include up-to-date standards that reflect current best practices and technologies—and to serve as a higher bar, from which the regulated community can continue to improve.

Finally, PHMSA published a Notice of Proposed Rulemaking (NPRM) in October 2020 for Class Location Change Requirements. As you are aware, PHMSA is required by statute to hold advisory committee meetings on our proposed rules to solicit recommendations to ensure our rulemakings are reasonable, feasible, cost-effective, and practicable. These advisory committee meetings have helped derive consensus around highly technical regulatory policies. The PIPES Act of 2020 requires PHMSA to hold an advisory committee meeting on the NPRM for this rule. Based on the aggressive timelines and safety priorities included in the PIPES Act of 2020, PHMSA is working to make efficient use of agency resources and advisory committee members' time and anticipates holding an advisory committee meeting on the NPRM in conjunction with publishing the most important safety agenda items identified by Congress.

It is important to point out that rulemaking is designed to be an iterative process that encourages maximum participation by all stakeholders and rigorous analysis in support of decision making. This process helps ensure the promulgation of comprehensive rules that protect the public and the environment and meet our statutory requirement for rules with benefits that exceed their costs. PHMSA holds public meetings and workshops and conducts significant outreach prior to rulemakings, using the information gathered to establish a legal record and to strive to craft the most effective rules possible. Such collaboration, well in advance of the rulemaking process, allows PHMSA to identify concerns and potential solutions and to allocate its limited resources where they are needed most. In the past, these comprehensive efforts have also helped avoid expending additional resources on legal challenges. In addition to congressionally mandated rules, many of PHMSA's rulemakings underway address important recommendations from the National Transportation Safety Board, resulting from safety issues identified during investigations in the aftermath of some tragic accidents. PHMSA's rules also address recommendations from the U.S. Government Accountability Office (GAO), the DOT Inspector General (DOT IG), and the agency's own safety findings. When PHMSA proceeds with such rulemakings identified by independent sources, it must make sure that its regulations account for known safety issues, technological feasibility, and cost-effectiveness.

Increased Litigation

With all the good work that is being done to advance pipeline safety by the promulgation of new rules, PHMSA also faces a new normal in terms of increased challenges to its rulemakings, resulting in longer development timelines and diversion of personnel resources to respond to legal challenges—which could otherwise be utilized to advance the myriad congressional directives and regulatory priorities of the agency and stakeholders.

PHMSA has also seen a dramatic increase in interest in its rulemakings pertaining to energy resources. By way of example, PHMSA's LNG by Rail Suspension NPRM, issued in November 2021, has had over 7,000 comments—including a coordinated letter-writing campaign by environmental advocacy organizations and a letter signed by over 20 State Attorneys General, as well as many members of the House of Representatives on both sides of the aisle.

Specifically, PHMSA has finalized four major rulemakings over the last year, and each of which has been the subject of judicial and/or administrative challenges. PHMSA currently faces pending litigation brought by pipeline industry trade groups on the Gas Gathering Final Rule, Valve Installation and Repair Final Rule, among others, from stakeholder groups and governments, across the spectrum.

PHMSA rulemaking resources are consequently spread thin. The same subject matter experts, attorneys, and economists who develop new PHMSA rules are also the ones who must help develop the briefs and arguments to respond to legal challenges after issuance.

Enforcement and Compliance

While PHMSA's enforcement cases have remained relatively steady, we have set records for our proposed civil penalties in 2021 and again in 2022. These cases, many of which are still being adjudicated, include the worst carbon dioxide pipeline incident on record as well as a case related to the 2021 Colonial Pipeline cybersecurity incident. PHMSA continues to pursue 100% collections of the civil penalties it has imposed; however, some operators with smaller civil penalties have significantly delayed paying the penalties they owe.

In terms of forward-looking, potential rulemakings, the PSRs currently include emergency planning, response, and timely notification requirements for pipeline operators. However, incidents involving, for example, carbon dioxide pipelines as well as LNG facilities, do not necessarily require communications to communities who rightfully have an increased fear of these facilities after a safety incident occurs. PHMSA will look to its rulemaking authorities to help address the lack of post-incident communications but we also welcome congressional ideas in this space.

Another issue we're examining involves the safety and performance of pipes manufactured outside of the U.S. Many larger operators deploy their own inspectors when utilizing foreign-made pipe in their projects—in order to ensure maximum safety and performance. However, when those U.S. companies find non-spec pipe (pipe not meeting Federal or industry standards), they may simply refuse to purchase it—which may result in another U.S. pipeline construction company ultimately purchasing or utilizing the same non-spec piping. On the hazardous materials side of our agency, PHMSA deploys inspectors across the globe to ensure products that are moving hazardous materials in the U.S. are inspected by U.S. inspectors. PHMSA is conducting analysis to better understand if non-spec foreign made pipes are being utilized in the U.S.

Research and Innovation

While PHMSA continues to advance pipeline safety by strengthening its regulations and enhancing its inspector training, inspections, and enforcement programs, many of the root causes of incidents are best addressed through research and technological innovation.

PHMSA's Pipeline Safety Research Program works with academia, the regulated community, private research consortiums and federal partners to sponsor research and development (R&D) projects focused on providing near-term solutions for pipeline transportation infrastructure issues that will improve safety, reduce environmental impact, and enhance reliability.

Hydrogen/Carbon Dioxide (CO2)

In FY 2022, PHMSA awarded approximately $6 million in research investments on hydrogen projects. Specifically, under the Competitive Academic Agreement Program, PHMSA awarded two projects on pipeline infrastructure and modernization for hydrogen networks to two universities. These research opportunities expose students to the pipeline safety sector to encourage them to join the federal or state pipeline safety workforce or the private sector after graduation. PHMSA also leverages the University Transportation Centers Program to meet its research needs.

Also, in FY 2022, PHMSA awarded four projects related to hydrogen pipelines and storage, under our Core Research Program, totaling just over $2 million. These projects will research the safe transportation and storage of hydrogen via repurposing existing infrastructure used for natural gas transport and underground storage, improving hydrogen leak detection, and characterizing hydrogen-specific pipeline integrity threats.

PHMSA is collaborating with the Department of Energy (DOE) and other DOT modes on developing a Memorandum of Understanding to establish collaborative partnerships on R&D and safety associated with the transport of carbon dioxide and hydrogen via pipelines, rail, barge, ship, and truck.

Lastly, PHMSA announced last year new research topics to better determine impact areas for the safer operations of carbon dioxide pipelines. The results of this may help inform a current rulemaking related to carbon dioxide pipelines but congressional attention related to these issues is also welcome.

LNG

Recent global fluctuations in natural gas supplies as well as a transition from more carbon-intensive energy sources continue to spark investments in LNG. Currently, there are eight LNG export terminals with a total LNG production capacity of approximately 14 billion standard cubic feet per day (bcf/d) in the United States.

There are also five LNG projects under construction, which will add an estimated 11.9 bcf/d in LNG production capacity. To that end PHMSA has funded 14 R&D LNG-safety-related research projects since 2007; with nine completed/closed and five currently active, all totaling $5.7 million.

Additionally, in the Consolidated Appropriations Act, 2023, Congress allocated up to $8.4 million to PHMSA for the creation of an LNG Center of Excellence aimed at positioning the United States as the leader and foremost expert in LNG operations—including safety and environmental performance. PHMSA has already initiated planning for the establishment of the Center which will enhance U.S. LNG operations and safety education and oversight and may result in LNG regulatory improvements. It will also serve as a repository of information and facilitate collaboration among stakeholders to enhance safety and environmental performance through research.

Funding for State Pipeline Safety Programs

Since 1970, when a national, uniform standard of pipeline safety regulations was published, states have had the authority, through PHMSA, to regulate the safety of intrastate pipelines. Under the authority of Sections 60105 and 60106 of Title 49U.S. Code (49 U.S.C.) for state pipeline safety program certifications, states have been allowed to assume safety authority for the inspection and enforcement of intrastate pipelines. PHMSA sets the minimum Federal regulations for pipeline safety, which the participating states then adopt into their state code and enforce. States are allowed, under 49 U.S.C. Section 60104(c), to adopt more stringent safety standards than the minimum standards PHMSA sets. This allows states to codify and enforce regulations that deal with specific, regional (or local) risks that might not be feasible or cost-beneficial to regulate at the Federal level. Many states have established safety regulations that are more stringent than the Federal regulations.

PHMSA relies on this extremely important partnership to accomplish its safety mission, which is to protect people and the environment by advancing the safe transportation of energy and other hazardous materials that are essential to our daily lives. New pipeline safety regulations and new infrastructure (such as gas gathering lines) specific to state safety authority have and will continue to require state pipeline safety programs to increase staff—in order to handle the additional infrastructure oversight responsibilities. These state pipeline safety programs employ approximately 435 inspectors who are responsible for inspecting over 85 percent of the Nation's pipeline infrastructure through certification with PHMSA.

The PIPES Act of 2020 allows PHMSA to pay not more than 80 percent of the total cost of the personnel, equipment, and activities reasonably required by the state agency for the conduct of its pipeline safety program during a given calendar year. This was changed from 50 percent in the 2006 PIPES Act. However, for fiscal years 2019 to 2021 State Base Grant federal funding covered less than 70 percent of the actual total state program costs. The actual federal funding is estimated to be approximately 60 percent of the state program's total costs for fiscal year 2022. Unfortunately, current year 2023 federal funding is estimated to be only 56 percent of the total state program costs—due, in part, to increasing needs from states. This is another area where we welcome congressional ideas on how to support states and their vital role in implementing many of the new regulations previously discussed.

Control Room Management and Cyber Security

Not only is the industry facing expansion in the number of regulated pipeline miles and changes in product demand, both industry and regulators are addressing the growing threat of cyberattacks. PHMSA, the Transportation Security Administration (TSA), and DOE have a mutual interest in ensuring coordinated, consistent, and effective activities that improve transportation security. PHMSA's safety oversight of pipeline control rooms forms a nexus with TSA's cybersecurity oversight, the Cybersecurity and Infrastructure Security Agency's (CISA's) role as the national coordinator for critical infrastructure security, and DOE's national energy management. The 2021 cyber-attack on Colonial Pipeline demonstrated how critical it is for a whole-of-government approach to safeguarding our Nation's critical infrastructure—as well as collaboration with the private sector when it comes to planning and communications.

PHMSA is leveraging its authorities to inspect and enforce three components of pipeline operations including pipeline control room regulations, integrity management plan requirements, and emergency response plan regulations by incorporating cybersecurity questions in inspections that focus on considering cyber as a risk and having emergency response plans in place that consider the threat of cyberattacks as well as envision measures to mitigate impacts to operations. PHMSA has also engaged with CISA and TSA on cybersecurity exercises for pipeline operators.

To help advance our safety mission as it relates to cybersecurity, PHMSA is in the process of hiring pipeline and cybersecurity experts to assist in control room management inspections and provide cyber expertise support to PHMSA leadership. Cyber specialists can better identify pipeline infrastructure-related cyber risks, incidents, and significant issues that PHMSA currently lacks expertise in. This would help increase information sharing, create better understanding between agencies, and identify key issues between the nexus point of safety and security.

PHMSA is increasing cybersecurity training opportunities for its staff, as well as the staff of its state partners. By expanding the knowledge base of inspectors, we are better positioned to identify risks during routine control room inspections and coordinate when needed with colleagues in TSA.

PHMSA is also developing an adaptable emergency preparedness plan that will address the Agency's response to all-threat, all-hazard, notice and no-notice incidents, including cybersecurity. As part of the process, PHMSA is developing an organizational framework to respond to those incidents, formal situational reporting internal and external to PHMSA, formal information sharing processes, and specific coordination methods between PHMSA, TSA, CISA, DOT Operating Administrations, DOE, and the Federal Emergency Management Agency.

OIG Inspection, GAO Audits, and NTSB Recommendations

In terms of PHMSA's compliance and inspection program, we recently underwent a DOT IG audit, which was initiated in May 2022, to review PHMSA's implementation of its Integrated Inspection Program. Throughout the audit, the PHMSA team provided detailed overviews and walkthroughs of its Integrated Inspection Program, including, but not limited to planning, training, inspection conduct, and governing policies. PHMSA even organized and facilitated the OIG's participation in several ongoing integrated inspections, and, at PHMSA's invitation, OIG personnel attended the Office of Pipeline Safety's annual inspection planning meeting in October 2022. The DOT IG provided helpful feedback and we consider the audit to have been beneficial to PHMSA's continual improvement. PHMSA learned valuable lessons and received three DOT IG recommendations. We continue to have a constructive working relationship with the DOT IG, and the audit helped us to continue to move toward our common goal of advancing pipeline safety.

In August 2020, the GAO published a study finding Federal agencies have incorporated most but not all key collaboration practices in the permitting processes for export facilities for LNG. The GAO identified key practices for PHMSA, that can help sustain collaboration among federal agencies.

PHMSA's team worked with GAO to adopt a process for conducting standards-specific reviews approximately every two to three years. The new process will ensure that a sufficient review is conducted and that PHMSA makes appropriate determinations about whether to update standards.

When it comes to our work with the National Transportation Safety Board (NTSB), PHMSA is addressing recommendations that include requiring control room operators to notify emergency call centers in impacted communities when potential ruptures take place, equip control rooms with supervisory controls and data acquisition systems to pinpoint leaks along transmission lines, and amend Title 49, Code of Federal Regulations (CFR) to require automatic shutoff valves or remote-control valves at high consequence areas. We continue working to resolve any open recommendations—some of which we are constrained by resources and some by statutes—such as the congressional prohibition on applying the Automatic/Remote Shut Off Valve Rule to existing pipelines.

PHMSA continues to work with NTSB to address recommendations that have been made following other natural gas and hazardous liquid accidents. We collaborate with NTSB often, including opportunities for cross-training of our respective staff. We'll continue to engage with NTSB as a partner in advancing safety.

All of these efforts are important because continual improvement is a key principle of safety management systems and high-reliability organizations, and one we embrace for both the agency and the industries we regulate.

Transparency, Equity, Environmental Justice, and Outreach to Underserved Communities

To both implement the President's executive orders on equity (EO 13985 and EO 14008), as well as to help address historic inequities in the transportation system, PHMSA's Office of Pipeline Safety has expanded its efforts to make public pipeline safety incidents and enforcement data (which was also recently the subject of a GAO report that lauded PHMSA's transparency and encouraged further actions). Specifically, PHMSA has created a publicly available pilot, interactive mapping tool that allows users to view the location of pipeline incidents, as well as a geographic overlay with underserved communities.

When PHMSA first viewed the preliminary information from this tool, staff felt inspired to act—to help ensure all communities are receiving requisite safety protections. As part of this effort, PHMSA has engaged our state and federal partners, as well as stakeholders, to share our findings, and they, too, are engaging in dialogues with pipeline operators to ensure maintenance and safety measures do not leave underserved communities behind.

These communities are identified through U.S. Census and internal DOT/PHMSA data focused on underserved and transportation-disadvantaged communities that have experienced excavation damages, and other pipeline incidents and accidents. PHMSA has also expanded its public outreach and education on pipeline awareness and safety as well as community-based excavation damage prevention initiatives to historically underserved and socioeconomically challenged geographic areas.

Increased Engagement with the Public

PHMSA is committed to enhancing all stakeholder engagement and has increased the number of public meetings and information briefings it hosts—holding four public meetings and information briefings so far in FY 2023, with additional public meetings and information briefings planned. Personally, I have visited community members and victims, on-site, where pipeline facilities have failed (e.g. Kalamazoo, MI; Bellingham, WA; Satartia, MS; and Freeport, TX).

PHMSA has also increased its engagement with public interest groups like the Pipeline Safety Trust, pipeline worker labor unions, and environmental groups, actively participating in conferences and meetings to hold a two-way dialogue on important pipeline safety issues, emphasizing that pipeline safety is a shared responsibility.

In November and December 2022, PHMSA partnered with the DOE in a series of Community Engagement Workshops on Carbon Capture, Utilization, and Storage and continues to serve as a resource regarding pipelines to DOE and the public. PHMSA has also supported requests from individuals and groups to participate in meetings to discuss CO2 pipeline projects to listen to concerns on safety, environmental justice, environmental impacts, and emergency response preparedness, as well as meet with representatives at the state legislature level. As previously noted, in May of 2022, I personally visited the community of Satartia, MS—about an hour northwest of Jackson—the site of one of the worst carbon dioxide pipeline incidents in history, in order to hear directly from the community and first responders that helped the community during that serious incident.

Similarly, just yesterday I visited with members of the community in and around Freeport, TX, which is home to the Freeport LNG facility. In June of 2022, an explosion at the facility resulted in a massive fireball and understandably left lasting concerns with the community. In February of this year, PHMSA, along with our coregulators of LNG export facilities (the Federal Energy Regulatory Commission and the U.S. Coast Guard) held a town hall meeting in the community—with simultaneous bilingual translation—to help inform the surrounding community members of our work to investigate the incident and require changes needed to enhance safety at the facility.

In 2022, PHMSA's Office of Pipeline Safety participated in nearly 220 public meetings, events, and conferences to educate our stakeholders on pipeline safety and damage prevention initiatives and to address questions about the Federal pipeline safety regulations or concerns about pipeline-related matters. PHMSA continues to promote the 'Call 811 Program' through participation in events as well as through social media and digital campaigns encouraging safe digging practices.

Efficiencies in Oversight, Taxpayer Stewardship, and Focus on Employees

Roughly 169 midstream oil and gas industry projects are expected to begin operations in the United States from 2021 to 2025, according to the Pipeline and Gas Journal. Over the last five years, liquid pipeline incidents have fallen by 21% while pipeline mileage and barrels delivered have increased by more than 27%. As previously noted, and to put it simply, our oversight responsibilities continue to grow both in terms of the types of facilities we regulate as well as the number of facilities we regulate: PHMSA has increasing responsibility for LNG facilities, underground natural gas storage, as well as natural gas gathering lines. PHMSA's budget, excluding the new gas distribution grant program, does not grow at a rate commensurate with its responsibilities. Consequently, PHMSA has had to continuously operate relatively leaner as compared to our expanded universe of regulated facilities. To this end, PHMSA has also utilized advisory bulletins, public meetings, research solicitations, and increased collaboration with coregulators such as the Federal Energy Regulatory Commission, the Environmental Protection Agency, the Department of Interior, the U.S. Coast Guard, and our state partners through collaboration with the National Association of Pipeline Safety Representatives.

Hiring times at PHMSA have been reduced by 25% unfortunately—due in part to the pandemic—PHMSA was not spared the so-called great resignation, losing many individuals to both retirements and other departures. But PHMSA is exploring ways to continue to improve the agency's hiring and recruitment to make it both more efficient and effective in recruiting and retaining talented applicants.

On the hiring, recruitment, and retention front, unfortunately—due in part to the pandemic—PHMSA was not spared the so-called great resignation, losing many individuals to both retirements and other departures.

To meet congressional directives to improve efforts to attract and retain pipeline engineers and inspectors, PHMSA has undertaken new recruitment and retention efforts—seeking approval from the Office of Personnel Management to increase special pay rates for some engineer inspectors—commensurate with similar federal special pay rates, developing new tuition reimbursement efforts, and utilizing new online recruitment methods. PHMSA is also utilizing the Department of Defense's Operation Warfighter (OWF) program that matches qualified wounded, ill, and injured Service members with federal internships for veterans to gain valuable work experience during their recovery and rehabilitation—and create a pathway from the military to permanent employment. PHMSA has kept up with the PIPES Act of 2020 hiring mandates—both for inspectors as well as for regulatory personnel, that have helped lead the agency to some of its most productive years ever in terms of both finalizing regulations as well as enforcement actions and a reduction trend in hazardous materials and pipeline incidents.

PHMSA has also utilized technologies like iPads to eliminate paperwork for inspectors—which has resulted in more efficient use of inspectors' time and increased the accuracy and standardization of inspections.

On an agency-wide basis, PHMSA has reduced or eliminated its use of nearly two dozen disparate software systems in favor of less costly, integrated systems. PHMSA is utilizing the cost savings of this nature to continue investing in more long-term, cost-saving programs.

Conclusion

In closing, I would like to thank you again for the opportunity to engage with you on the critical issues facing PHMSA and in turn facing a major component of the largest, most sophisticated energy transportation system in the world. And most importantly, I would like to emphasize my deep gratitude to the nearly 600 full-time federal employees and nearly 200 contractors that make up what I believe is the most unsung agency in the Federal Government. Congress has charged us with tremendous responsibilities—from ensuring the safe transportation of some of the most valuable goods that move in commerce, like satellites and spacecraft, as well as some of the most essential goods like fertilizer used on our farms, which can be transported by pipeline. As we take on ever greater oversight responsibilities with oversight of the build-out of carbon dioxide and hydrogen pipelines and other energy products of the future, PHMSA must either continue to grow our resources, or continue to reassess multiple and increasing priorities with the same amount of resources.

We look forward to continuing to work with Congress to improve pipeline and hazardous materials safety and to reduce associated environmental impacts.

Thank you again for inviting me here today. I look forward to your questions.

Mr. NEHLS. Thank you, Mr. Brown.

Mr. Black, you are now recognized for 5 minutes.

Testimony of Andrew J. Black, President and Chief Executive Officer, Liquid Energy Pipeline Association (LEPA)

Mr. BLACK. Thank you, Mr. Chairman, Ranking Member, and members of the subcommittee. I am Andy Black, president and CEO of the Liquid Energy Pipeline Association.

LEPA represents pipeline owners and operators delivering transportation fuels like gasoline, diesel, and jet fuel; transportation feedstocks like crude oil; home heating fuels like propane; industrial feedstocks like ethane; and low-carbon solutions like renewable diesel, carbon dioxide, and liquefied petroleum gas. We have over 50 members delivering over 20 billion barrels annually across a pipeline network of nearly 230,000 miles.

As Congress examines Federal pipeline safety, let me begin by providing some data and information on the current state of liquid pipeline safety. According to publicly available data from PHMSA, pipelines are the safest way to transport energy, and they are getting safer. Over the last 5 years, total incidents from liquid pipelines have dropped 25 percent. Incidents impacting people and the environment are down 15 percent. This last metric, incidents impacting people and the environment, was developed jointly by PHMSA, the Pipeline Safety Trust, and industry under the recommendation of the National Transportation Safety Board.

Not only are pipelines getting safer, they are the safest way to deliver energy. A report for Congress completed by PHMSA compared incident and spill rates across pipeline, rail, and trucks. PHMSA found pipelines have a lower incident rate per barrel of crude oil shipped compared to rail or truck, and the lowest percentage of crude oil released compared to rail or truck.

Administration analysis and conclusion that pipelines are the safest way to deliver energy is also bipartisan. Secretary Hillary Clinton's State Department found not only would a major proposed pipeline be the safest way to deliver energy, it would be safer and have less impact on the environment than taking no action at all. The Obama administration found rejecting that pipeline project and shipping the same energy by rail would increase the risk of oil release by over 800 times, and barrels released by 2.6 times. In many locations, shipment by rail or truck is necessary and generally a safe delivery mode. But we are proud that pipelines are the safest way to deliver the energy American consumers need and want.

We are also not resting on improving pipeline safety. In partnership with the American Petroleum Institute, we have ongoing initiatives to improve public engagement, develop best practices for managing geohazards and seismicity, improve pipeline inspection technologies, implement pipeline safety management systems, and share safety improvement lessons from pipeline incidents and near misses. Before that, in the last several years, we have completed initiatives on crack management, pipeline integrity management, data integration, hydro testing, and emergency response.

We look forward to Congress reauthorizing Federal pipeline safety provisions and making improvements to PHMSA pipeline programs. LEPA recommends improvements in three categories: leveraging safety technologies and knowledge; safe, low-carbon future; and improving PHMSA safety programs. I have attached to my testimony brief summaries of each of our recommendations; I will highlight two today.

First, a continuing frustration for pipeline operators is how old are some of PHMSA's regulations. Key requirements for inspecting and repairing pipelines are now over 20 years old. The pipeline technologies and analytic methods on which they are based are woefully out of date, replaced by proven technology and analytics and engineering methods that PHMSA should reflect in its regulations, but often does not.

Congress in the 2020 PIPES Act, section 104, authorized a pipeline safety technology demonstration program to demonstrate cutting-edge pipeline safety technologies and advanced analytics. Thank you, Congress. But PHMSA, in implementing the program, added a number of administrative hurdles and requirements beyond what Congress mandated, making the program untenable. We have proposals to unlock this program and others from unnecessary Government redtape.

Second, the pipeline industry supports updating regulatory requirements for low-carbon solutions. A major build-out of CO2 pipelines is necessary to transport CO2 from where it is captured to where it can be stored permanently out of harm's way. Without a new network of CO2 pipelines, we will not meet our goals for reducing greenhouse gas emissions.

There are already dozens of Federal regulatory requirements administered by PHMSA covering CO2 pipeline safety, design, construction, operation, maintenance, pressure testing, and more. In a handful of discrete areas, PHMSA requirements would benefit from updates reflecting the latest safety approaches and learnings. We believe a targeted approach is the best way for PHMSA to issue new requirements quickly. This would also help PHMSA to avoid getting bogged down in an open-ended exercise that, like other PHMSA rulemakings, can take many years.

I am happy to answer any questions, and I thank the subcommittee for the opportunity to testify.

[Mr. Black's prepared statement follows:]

Prepared Statement of Andrew J. Black, President and Chief Executive Officer, Liquid Energy Pipeline Association (LEPA)

Thank you, Chair, Ranking Member, and members of the subcommittee. My name is Andy Black and I am President and CEO of the Liquid Energy Pipeline Association. LEPA represents pipeline owners and operators delivering transportation fuels like gasoline, diesel, and jet fuel, transportation feedstocks like crude oil, home heating fuels like propane and home heating oil, industrial feedstocks like ethane and butane, and low carbon solutions like renewable diesel, liquified petroleum gas and carbon dioxide. We have over 50 member companies delivering over 20 billion barrels annually across a nearly a 230,000-mile network of pipelines.

As Congress examines federal pipeline safety provisions and programs, let me first begin by providing some data and information on the current state of liquids pipeline safety. According to publicly available government data from PHMSA, pipelines are the safest way to transport energy and they are getting safer.

Over the last 5 years, total incidents from liquids pipelines have dropped over 25%. Incidents impacting people or the environment are down 15%. This last metric, Incidents Impacting People or the Environment, was developed jointly by PHMSA, the Pipeline Safety Trust and industry under the recommendation of the National Transportation Safety Board. NTSB asked the pipeline community to identify the most meaningful metric for measuring pipeline safety. You'll hear from witnesses about various metrics and PHMSA certainly tracks many. But we agree that Incidents Impacting People or the Environment are the most meaningful and are gratified they are down 15% over the last 5 years.

Not only are pipelines getting safer, they are the safest way to deliver energy. A 2018 report completed by PHMSA at the direction of the Senate Appropriations Committee[77] compared incident and spill rates across pipeline, rail and trucks. PHMSA found pipelines have a lower incident rate per barrel of crude oil shipped compared to rail or truck. PHMSA also found pipelines result in the lowest percentage of crude oil released compared to rail or truck.

[77] Report on Delivering Crude Oil by Truck, Rail, and Pipeline, Office of Hazardous Materials Safety, U.S. Pipeline and Hazardous Materials Safety Administration, 2018.

Administration analysis and conclusion that pipelines are the safest way to deliver energy is also bipartisan. Secretary Hillary Clinton's State Department during the Obama administration[78] found not only would a major proposed pipeline be the safest way to deliver energy, it would be safer and have less impact on the environment than taking no action at all. The Obama administration found rejecting that pipeline project and shipping the same energy by rail increased the risk of oil release by over 800 times and barrels released by 2.6 times.

We certainly understand in many locations shipment by rail or truck is necessary and generally a safe delivery mode on their own. But we are proud that pipelines are the safest way to deliver the energy American consumers need and want.

We also are not resting on improving pipeline safety. The pipeline industry wants to reduce the number of incidents and barrels released even further. That is why we partner with the American Petroleum Institute and our member companies on numerous industry-wide initiatives to improve pipeline safety.

We have ongoing industry-wide initiatives to improve public engagement, develop best practices for managing geohazards and seismicity, improve pipeline inspection technologies, implement pipeline safety management systems, and share safety improvement lessons from pipeline incidents and near misses across our pipeline community. In the last several years, we have completed initiatives on crack management, pipeline integrity management, data integration, hydrotesting and emergency response. Our upcoming API–LEPA strategic plan for pipeline excellence will include new goals on cybersecurity, attracting, training and retaining quality personnel, and low carbon solutions. We will release that in May and will look forward to briefing the committee and member offices on its safety programs.

We also look forward to Congress reauthorizing federal pipeline safety provisions and making improvements to PHMSA pipeline programs. LEPA recommends pipeline safety improvements in three categories: 1) Leveraging Safety Technology and Knowledge, 2) Safe Low Carbon Future, and 3) Improving PHMSA Safety Programs. I have attached to my testimony further brief summaries of each of our specific recommendations and so will highlight just a few here today. A continuing frustration for pipeline operators is how old are some of PHMSA's regulations. Key requirements for inspecting and

[78] Final Supplemental Environmental Impact Statement for the Keystone XL Project, Bureau of Oceans and International Environmental and Scientific Affairs, U.S. Department of State, 2014.

repairing pipelines are now over 20 years old. The pipeline technologies and analytic methods on which they were based are woefully out of date, replaced by new technologies and analytic methods that PHMSA should reflect in its regulations but often does not. Congress in the 2020 PIPES Act authorized a program to demonstrate cutting edge pipeline safety technologies and advanced analytics. The hope is data from these technology pilots would help PHMSA modernize their regulations. However, PHMSA in implementing the program added a number of additional administrative hurdles and requirements beyond what Congress mandated, making the program untenable. We have proposals to unlock this program and others from unnecessary government red tape.

The pipeline industry also supports updating regulatory requirements for low carbon solutions, such as carbon dioxide pipelines. A major buildout of CO2 pipelines is necessary to transport CO2 from where it is captured to where we will store it permanently out of harm's way. Without a new network of CO2 pipelines, we will not meet goals for reducing greenhouse gas emissions. We want policymakers and the public to know these CO2 pipeline systems will be covered by federal pipeline safety requirements. Many do not know there are already dozens of federal regulatory requirements administered by PHMSA covering CO2 pipeline safety. However, we agree that in a handful of discreet areas, PHMSA requirements would benefit from updates reflecting the latest safety approaches and learnings. We also believe a targeted approach is the best way for PHMSA to issue new requirements quickly. This would also help PHMSA avoid getting bogged down in an open-ended exercise that like other recent PHMSA rulemakings could take many years.

We also support PHMSA having the expert pipeline safety personnel it needs to complete its mission. Congress often places new regulatory mandates on PHMSA. Similarly, lessons learned from incidents, advances in technology, or our changing infrastructure systems can drive the need for PHMSA to issue guidance or undertake rulemakings. We encourage Congress to help PHMSA get the experts on pipeline safety it needs to meet the expectations of Congress and the public.

On this or any of our proposals or those under consideration by the subcommittee I'm happy to answer questions and thank the subcommittee again for inviting me to testify today.

ATTACHMENT

Legislative Priorities

2023 Pipeline Safety Reauthorization

The Liquid Energy Pipeline Association (LEPA) represents over 50 pipeline operators with nearly 230,000 miles of pipelines across America delivering affordable, reliable and plentiful energy to American drivers, families, farmers, workers and shoppers.

LEPA members deliver transportation fuels like gasoline, diesel and jet fuel, agriculture and rural home heating fuels like propane, heating fuels for the North-eastern U.S. like fuel oil and ultra low sulfur diesel, industrial feedstocks like ethane and butane, transportation fuel feedstocks like crude oil, and low carbon solutions like renewable diesel, ethanol, liquified petroleum gas and carbon dioxide. LEPA urges Congress to consider the following as they review federal pipeline safety laws in 2023:

Leverage Safety Technology & Knowledge

Hi-tech inspection and analytical tools, like an MRI or ultrasound in the doctor's office, are available for pipeline safety. However, key parts of PHMSA safety regulations are over 20 years old and do not reflect the latest advances in technology or know-how. Congress can do more to help modernize pipeline safety programs.

- Improve Pipeline Safety Technology Demonstration Program
- Promote Safety Sharing through Voluntary Information Systems
- Modernize Pipeline Incident Notification with Online Filing
- Allow Risk-Based Inspections for Storage Tanks
- Incorporate Leading Safety Standards into Program Requirements

Safe Low Carbon Future

Pipelines are the safest way to deliver liquid products, including low carbon economy byproducts like carbon dioxide. While existing federal regulatory requirements on CO2 pipelines are extensive, Congress can provide additional safety measures and clarifications to close remaining gaps in pipeline safety programs.

- Update CO2 Pipeline Safety Standards
- Clarify Methane Regulation Scope

Improve PHMSA Safety Programs

PHMSA, like the pipeline operators it oversees, can also benefit from adjustments reflecting continuous improvement efforts. Congress can help PHMSA increase the effectiveness and transparency of its pipeline safety programs and requirements.

- Optimize PHMSA Valve Inspection Scope
- Increase PHMSA Inspection Program Transparency
- Focus PHMSA Special Permit Program
- Improve PHMSA Enforcement Processes
- Provide Expert Pipeline Safety Regulatory Personnel
- Enhance Pipeline Right of Way Management
- Jumpstart Required Regulation of Idled Pipe
- Close Loophole in Penalties for Pipeline Attacks

Improve Pipeline Safety Technology Demonstration Program

Action Needed:

Congress should prevent bureaucratic red tape from stifling new technologies and analytics that could improve pipeline safety.

Background:

In the 2020 PIPES Act, Congress recognized pipeline safety could benefit from harnessing the latest hi-tech inspection technologies and analytics. Congress authorized PHMSA to conduct a pipeline safety technology demonstration pilot program under certain conditions. However, PHMSA added a host of additional administrative, regulatory and legal conditions to the program, effectively preventing its use.

Status:

PHMSA has received no applications to conduct technology pilots. Pipeline operators cite the additional conditions PHMSA imposed in its implementation guidance as making the program infeasible.

Promote Safety Sharing through Voluntary Information Systems

Action Needed:

Congress should authorize a Voluntary Information Sharing (VIS) program based on the recommendations of the public advisory committee formed pursuant to the 2016 pipeline safety reauthorization law.

Background:

Programs where regulators, operators, vendors, unions and safety advocates can share information, with appropriate legal protections, and develop joint safety recommendations, have found success at FAA. Interested stakeholders developed a legislative proposal to implement recommendations of the Congressional VIS public advisory committee. That language did not reach consensus before the 2020 Pipes Act.

Status:

Stakeholders appear to have agreement on draft legislative language after overcoming remaining issues which prevented VIS inclusion in the 2020 Pipes Act.

Modernize Pipeline Incident Notification with Online Filing

Action Needed:

Congress should require the National Response Center (NRC) to develop and allow use of a simple and quick online incident notification system.

Background:

The NRC currently requires placing a telephone call to the NRC to notify it of a pipeline incident. Operators and regulators have experienced lengthy delays waiting for a live person to answer the phone at NRC. In the decades since the NRC first established its telephone requirement, web-based applications accessed online now allow for simple and quick notification to all stakeholders.

Status:

Online notification of pipeline incidents, such as through a simple online app, is not currently available.

Allow Risk-Based Inspections for Storage Tanks

Action Needed:

Congress should prevent unnecessary greenhouse gas and air pollutant emissions, worker safety threats and hazardous waste by allowing risk-based storage tank inspections.

Background:

PHMSA regulation requires inspection of petroleum storage tanks on fixed schedules, regardless of their actual maintenance needs. The early draining and cleanout of tanks to perform unneeded inspections releases greenhouse gas emissions and air pollutants, subjects workers to unnecessary risks in confined spaces, and creates hazardous waste needing disposal. Current industry standards on tank inspection are already accepted by other federal and state agencies, including EPA's program.

Status:

Engineering assessments methods to set maintenance schedules that avoid climate, environmental and worker impacts are available for PHMSA to incorporate into regulation.

Incorporate Leading Safety Standards into Program Requirements

Action Needed:

Congress should direct PHMSA to review pipeline safety best practices and incorporate in a timely manner, where appropriate. Congress should also encourage PHMSA to participate more fully in the standard setting process.

Background:

Pipeline operators develop industry-wide best practices in an open and multi- stakeholder collaborative process certified by the American National Standards Institute. These practices are technical documents developed by engineers and PHMSA has incorporated several into their regulations.

Status:

PHMSA is often slow to incorporate the latest editions of best practice standards. LEPA's legislative proposal would require PHMSA to review the latest editions to regulated industry standards in a timely manner and incorporate them where appropriate through the regular notice and comment process.

Update CO2 Pipeline Safety Standards

Action Needed:

Congress should direct PHMSA to update existing carbon dioxide pipeline safety regulations in key areas.

Background:

Current federal pipeline safety requirements already regulate CO2 pipeline design, construction, operation, maintenance and emergency response. However, there are targeted areas in system coverage, impact modeling, maintenance and emergency response where PHMSA can do more to address the specific needs of CO2 pipelines. These additions will promote new low carbon infrastructure.

Status:

While PHMSA has announced general plans to impose additional CO2 regulation in the future, an opportunity exists to direct PHMSA to act more quickly in key areas that will improve CO2 pipeline safety.

Clarify PHMSA Methane Regulation Scope

Action Needed:

Congress should clarify the 2020 PIPES Act provisions covering methane monitoring and mitigation from natural gas pipelines does not apply to hazardous liquids pipelines.

Background:

The 2020 PIPES Act requires PHMSA to issue regulations for the monitoring and mitigation of methane from natural gas pipelines. Hazardous liquids pipelines delivering crude oil, gasoline, diesel fuel or other energy liquids do not transport or release methane and the PIPES Act does not include them. However, some PHMSA inspectors are attempting to require hazardous liquids pipeline operators to update their inspection and maintenance plans to meet methane monitoring and mitigation program requirements.

Status:

PHMSA interpretation and implementation of the 2020 PIPES Act methane provisions remains uneven.

Optimize PHMSA Valve Inspection Scope

Action Needed:

Congress should direct PHMSA to revise its recent pipeline rupture rule to only apply additional inspection requirements to valves related to pipeline ruptures, and drop misapplied inspection requirements of all valves, including those which would not contribute to or limit a rupture.

Background:

In 2022, PHMSA required semi-annual inspection of valves that could mitigate the extent of a pipeline rupture. However, between the proposed and final versions of the rule PHMSA without notice expanded the requirement to include all valves, even those unrelated to addressing ruptures. PHMSA claims this expansion was unintended but has not addressed the legal uncertainty created by the change.

Status:

Informal PHMSA discussions to issue guidance would not change the clear regulatory language and legal liability presented by the final regulation.

Increase PHMSA Inspection Program Transparency

Action Needed:

Congress should require PHMSA annually to share publicly its inspection priorities and number of federal and state inspections of pipelines across PHMSA regions. PHMSA should also ensure any guidance used by PHMSA inspectors is publicly available.

Background:

The current system of PHMSA region inspections and state inspection of pipelines can lead to unbalanced and duplicative inspection patterns across the country. More transparency would help ensure PHMSA's inspection program is meeting pipeline safety priorities.

Status:

While PHMSA is transparent with public sharing of its enforcement activities, PHMSA does not publicly share its inspection priorities or numbers of inspections across PHMSA regions and states.

Focus PHMSA Special Permit Program

Action Needed:

Congress should direct PHMSA to focus requirements of its special permit program on issues presented by the permit request and require timely review of applications.

Background:

Key PHMSA inspection and maintenance regulations are over 20 years old reflecting outdated technology and knowledge of the time. Federal law authorizes PHMSA to issue special permits varying from existing regulations as long they achieve equal levels of safety or are consistent with the public interest. However, PHMSA's special permit implementation program has grown over the years to include numerous requirements unrelated to the variances sought. Permit reviews can also take multiple years.

Status:

PHMSA's special permit process remains stuck in unnecessary requirements, reviews and delays.

Improve PHMSA Enforcement Processes

Action Needed:

Congress should direct PHMSA to establish a process when opening PHMSA enforcement proceedings to the public.

Background:

PHMSA historically conducted informal enforcement proceedings without public attendance. Public interest in pipeline cases is growing and PHMSA has allowed public attendance. PHMSA should set an orderly process for public attendance and protection of confidential information during proceedings.

Status:

There is no established process when opening PHMSA enforcement proceedings to the public.

Provide Expert Pipeline Safety Regulatory Personnel

Action Needed:

Congress should direct DOT and PHMSA to hire experienced pipeline safety regulatory personnel.

Background:

PHMSA is a small agency with personnel limitations that prevent it from timely addressing Congressional mandates and evolving pipeline safety needs. The 2020 PIPES Act authorized PHMSA to hire additional inspection and regulatory personnel. PHMSA used the resources to hire environmental policy experts and develop a methane monitoring and mitigation rulemaking. Work on other pipeline safety priorities face years of delay.

Status:

DOT human resources requirements placed on PHMSA limited hiring to junior personnel with less skill and experience. There are no requirements to hire pipeline safety expertise.

Enhance Pipeline Right of Way Management

Action Needed:

Congress should direct PHMSA to allow pipeline right of way conservation practices that benefit habitat and the environment.

Background:

PHMSA regulations requires activities along pipeline rights of way which can prevent natural habitat benefitting local wildlife and the environment. Conservation practices are available that balance pipeline monitoring and conservation.

Status:

Pipeline industry conservation initiatives are available for operator use if PHMSA allows the practice.

Jump Start Required Regulation of Idled Pipe

Action Needed:

Congress should direct PHMSA to complete the Congressionally mandated idled pipe rulemaking.

Background:

Operators sometimes suspend service on pipeline systems to reflect market conditions and temporary lack of demand for product movement along certain routes. However, no current PHMSA regulations govern maintenance and monitoring of pipelines in an idled status or the steps they must take to return to full service. Congress in the 2020 PIPES Act mandated PHMSA issue regulations addressing idled pipe.

Status:

The Congressionally mandated 2-year deadline for PHMSA to complete an idled pipeline rulemaking expired December 2022. PHMSA regulatory agenda currently predicts it will not issue a proposal before November 2023 with a future finalization date undetermined.

Close Loophole in Penalties for Pipeline Attacks

Action Needed:

Congress should close loophole in criminal penalties for pipeline attacks that are dangerous but do not result in damage.

Background:

Multiple past cases of attacks on pipelines and pipeline infrastructure posed a danger to the attackers, general public and environment. The federal government has obtained convictions on attacks resulting in physical damage to energy infrastructure. However, attacks that manipulate pipeline valves or other equipment, while not resulting in physical damage, are still dangerous due to risk of explosion or product release from a pressure buildup and rupture. Judges have dismissed prosecution of such cases because they did not result in physical damage, as the statute is currently interpreted.

Status:

Provisions exist in Title 18 against damaging energy infrastructure and Title 49 against damaging or destroying pipeline infrastructure, but there are no provisions against dangerous non-damaging activities.

Mr. NEHLS. Thank you, Mr. Black.

Mr. Grubb, you are now recognized for 5 minutes.

Testimony of Kenneth W. Grubb, Chief Operating Officer, Natural Gas Pipelines, Kinder Morgan, Inc., on Behalf of the Interstate Natural Gas Association of America (INGAA)

Mr. GRUBB. Chairman Nehls, Ranking Member Payne, and members of the subcommittee, good morning. My name is Kenneth Grubb, and I am the chief operating officer for natural gas pipelines at Kinder Morgan. As an engineer with over 30 years of experience in the pipeline sector, I appreciate the opportunity to testify before your subcommittee today.

Kinder Morgan is one of the largest energy infrastructure companies in North America. We have an interest in or operate approximately 82,000 miles of pipeline that transport natural gas and other products that are essential to daily lives.

Kinder Morgan is a member of the Interstate Natural Gas Association of America, also known as INGAA, which is a trade association that represents the interstate natural gas pipeline and storage industry. I am here today representing INGAA.

INGAA's members transport the vast majority of the natural gas consumed in the United States through a network of approximately 200,000 miles of interstate transmission pipelines. These transmission pipelines are critical infrastructure systems spanning multiple States, and deliver natural gas to end users such as local distribution companies, electricity generators, industrial manufacturers, and LNG export facilities.

Natural gas is the cleanest burning fossil fuel, and as demand for energy increases, expanded use of natural gas has helped improve air quality across the country by offsetting the use of higher carbon-intensive fuels. According to the Energy Information Administration, between 2005 and 2019, carbon dioxide emissions from the U.S. power sector declined by 33 percent, with natural gas accounting for more than half of those reductions.

Additionally, INGAA's members are committed to working together as an industry towards achieving net-zero greenhouse gas emissions by 2050. With that said, there are four key points I wish to make today.

First, INGAA's number-one priority is safety, and we support having a strong safety regulator. Regulators and industry experts alike, including the U.S. Department of Transportation and PHMSA, have agreed for decades that pipelines are the safest mode of energy transportation. Accidents are rare, and INGAA's members are committed to making them increasingly infrequent. INGAA's members work every day towards a goal of zero pipeline incidents.

To help achieve this goal, INGAA fundamentally believes in having a strong safety regulator, and we also support robust, durable regulations written by PHMSA to ensure the safety of our Nation's pipeline systems.

Second, PHMSA should complete its work on the class location rulemaking and issue a final rule. INGAA's top regulatory priority is completion of the class location rulemaking, which presents opportunities to increase safety and protect the environment. Existing regulations affecting this issue are over 50 years old. It is long overdue to update them to reflect modern safety technologies. Every year this regulation is not final, we estimate that 800 million standard cubic feet of natural gas is unnecessarily released to the atmosphere.

Additionally, INGAA members unnecessarily spend between $200 and $300 million per year to replace perfectly safe pipe. INGAA is hopeful that PHMSA will quickly finalize this crucial rule.

Third, the Gas Pipeline Advisory Committee, also known as GPAC, strengthens rulemakings and should meet more frequently.

The GPAC, which provides technical safety guidance to PHMSA, is comprised of equal representation from the natural gas industry, Federal and State agencies, and the public. Until recently, GPAC met frequently. However, since January 2021, the GPAC has only met once. With the known benefits of GPAC, INGAA believes that Congress should consider requiring PHMSA to hold at least two GPAC meetings per year.

Lastly, PHMSA should prioritize hiring safety engineers to assist with its rulemakings. INGAA recognizes that pipeline safety regulations are complicated and often take years to write. To help assist with rulemaking backlogs that frequently occur, PHMSA should hire more safety engineers who already understand pipeline systems. This will result in a more efficient process to produce effective rules.

In conclusion, I truly appreciate the opportunity to testify in front of your subcommittee today. Your efforts are critical to ensuring PHMSA has the resources and direction necessary to continually improve safety in our industry.

I look forward to your questions, and I thank you for your time. [Mr. Grubb's prepared statement follows:]

Prepared Statement of Kenneth W. Grubb, Chief Operating Officer, Natural Gas Pipelines, Kinder Morgan, Inc., on behalf of the Interstate Natural Gas Association of America (INGAA)

Chairman Nehls, Ranking Member Payne, and Members of the Subcommittee:

Good morning. My name is Kenneth Grubb, and I am the Chief Operating Officer for Natural Gas Pipelines at Kinder Morgan.

Kinder Morgan is one of the largest energy infrastructure companies in North America. Access to reliable, affordable energy is a critical component to improving lives around the world, and we are committed to providing energy transportation and storage services in a safe, efficient, and environmentally responsible manner for the benefit of people, communities, and businesses we serve. We have an interest in or operate approximately 82,000 miles of pipelines, 140 terminals, 700 billion cubic feet (Bcf) of working natural gas storage capacity and have renewable natural gas generation capacity of approximately 2.2 Bcf per year of gross production with up to an additional 5.2 Bcf in development. Our pipelines transport natural gas, renewable fuels, refined petroleum products, crude oil, condensate, CO2 and other products, and our terminals store and handle various commodities including gasoline, diesel fuel, chemicals, ethanol, metals and petroleum coke.

Kinder Morgan is a member of the Interstate Natural Gas Association of America (INGAA), which is a trade association that represents the interstate natural gas pipeline and storage industry. I am here today representing INGAA.

INGAA's members transport the vast majority of the natural gas consumed in the United States through a network of approximately 200,000 miles of interstate transmission pipelines. These transmission pipelines are analogous to the interstate highway system; in other words, they are large capacity, critical infrastructure systems spanning multiple states or regions to deliver natural gas to end users such as local distribution companies, electricity generators, industrial manufacturers and LNG export facilities.

I serve as Chief Operating Officer (COO) of the Natural Gas Pipelines Group for Kinder Morgan, Inc. In this capacity, I am responsible for all operational activities of the Kinder Morgan Natural Gas Pipelines Group, which encompasses approximately 70,000 miles of pipelines including natural gas transmission, gas storage, gathering, and processing facilities. I have served in this capacity since 2017. Before assuming this role, I held various other roles in operations, engineering, system design, project management, and construction and have over 32 years of experience in the energy sector.

For more than a decade, the shale revolution has gifted our country with abundant natural gas supplies, which has elevated the need for additional infrastructure to move it around the country. Pipelines make it possible to deliver North America's abundant natural gas reserves to fuel our homes, businesses, and the American economy, and are the safest and most efficient way to transport this critical energy source.

INGAA's members deliver clean, abundant, affordable natural gas throughout North America. Natural gas is the cleanest burning fossil fuel and, as demand for energy increases, expanded use of natural gas has helped improve air quality across the country by offsetting the use of higher carbon-intensive fuels. According to the Energy Information Administration, between 2005–2019, carbon dioxide emissions from the U.S. power sector declined by 33 percent, with natural gas accounting for more than half of those reductions. INGAA's members are committed to modernizing our nation's interstate natural gas delivery network infrastructure, lowering emissions from our operations, and mitigating the impacts of climate change by working together as an industry towards achieving net-zero greenhouse gas (GHG) emissions by 2050.

Thank you for the opportunity to testify at this hearing. There are four key points I wish to make in this testimony on behalf of the natural gas transmission pipeline industry.

1. INGAA's Number One Priority Is Safety, and We Support Having a Strong Safety Regulator

Regulators and industry experts alike, including the U.S. Department of Transportation and the Pipeline and Hazardous Materials Safety Administration (PHMSA), have agreed for decades that pipelines are the safest mode of energy transportation. Accidents are rare, and INGAA's members are committed to making them increasingly infrequent. INGAA's members work every day towards a goal of zero pipeline incidents.

INGAA fundamentally believes in having a strong safety regulator. For years, INGAA has sought robust, durable regulations led by PHMSA to ensure that all operators are held accountable to operate their systems in the safest manner possible. We take our commitment to safety seriously and appreciate the role that PHMSA plays to ensure that industry keeps its focus, and the public can have confidence in the safety and reliability of natural gas pipelines.

Pipeline companies consider safety every step of the way, from planning, to construction, to maintenance. Our members purchase top-quality materials, address any potential safety or security issues during the pipeline planning and citing process, and conduct consistent quality and safety checks throughout the construction process. Once operational, pipeline companies work to prevent releases by evaluating, inspecting, and maintaining pipelines.

Kinder Morgan is committed to the safe operations of our assets. Our pipeline and personal safety program metrics along with our safety goals are transparent to the public and are prominently posted on our website. We remain engaged with PHMSA and in-line inspection (ILI) tool vendors to further the development of ILI tool technology. As an example, one of Kinder Morgan's pipelines was the first to gain PHMSA acceptance to utilize Electro-Magnetic Acoustic Transducer (EMAT) technology to assess for the presence of environmentally induced cracking threats in lieu of the traditional hydrostatic testing technology that was used to manage these threats previously.

As part of on-going safety, pipeline companies conduct integrity management and continuous improvement programs in the areas of evaluation, inspection, and maintenance. A key component of integrity management programs is the use of ILI tools, sometimes called "smart pigs." Pipeline companies run these tools to detect any potentially harmful defects in the pipeline. These modern methods of pipe inspection have improved greatly over the last 30 years and are more effective, efficient, and environmentally sound compared to other assessment methods, with the added benefit of not significantly interrupting the operation of the pipeline.

INGAA's commitment to safety has been an essential priority for years. After the unfortunate and tragic incident in San Bruno, California, in 2010, INGAA member companies worked proactively to improve the industry's safety performance. This effort developed a set of guiding principles for pipeline safety, anchored around a goal of zero pipeline incidents, titled the "Integrity Management, Continuous Improvement" (IMCI) program. Since its inception, our industry has made rapid advances in safety technology and practices in continuous pursuit of our achieving this goal. INGAA members recently updated the IMCI program to ensure the reliability and resiliency of our infrastructure as we work to safely support the energy transition and evolve to a net-zero GHG economy. Similar to the EMAT technology previously mentioned, we also focused on advancing safety from newer technologies that will hopefully become more widespread throughout the industry and by regulators. This updated effort, titled IMCI 2.0 was created with the input of PHMSA, the National Transportation Safety Board, the National Association of Regulatory Utility Commissioners, the National Association of Pipeline Safety Representatives, and the Pipeline Safety Trust. The IMCI effort follows five guiding principles:

- Our goal is zero incidents;
- We are committed to a strong safety culture;
- We will be relentless in our pursuit of improving by learning;
- We are committed to implementing and continuously improving pipeline safety; management systems, and;
- We will regularly engage our stakeholders.

INGAA's work on the IMCI 2.0 program is nearly complete and we plan to share the results with key stakeholders later this year.

2. PHMSA Should Complete Its Work on the Class Location Rulemaking and Issue a Final Rule

INGAA's top regulatory priority with PHMSA is completion of the class location rulemaking, which presents opportunities to increase safety and protect the environment. The class location change regulations have not been substantively updated in over 50 years and revising them has been an INGAA goal for more than two decades. We were pleased that PHMSA issued a Notice of Proposed Rulemaking (NPRM) on the class location rule in October 2020. We were also greatly appreciative that Congress included a provision in the enacted 2020 PIPES Act that required the agency to hold a Gas Pipeline Advisory Committee (GPAC) meeting to review the NPRM by the end of 2021.

This proposed rulemaking would address scenarios where population changes around our pipelines necessitate changes to existing pipeline infrastructure. When a class location change occurs, the current regulations may require operators to replace the existing pipe even when an engineering assessment, including modern inspection tools, has shown it to be in safe, operational condition. The advancements in ILI tools and other safety technologies help enhance company decision making to make repairs and, in many cases, lessen the need for disruptive pipe replacements.

This causes two main problems. When PHMSA requires operators to replace pipes, operators must ensure that no gas is in the pipe they are about to replace, which results in service disruptions and natural gas being released to the atmosphere. Secondly, INGAA estimates that the existing requirements costs its members $200–$300 million per year to unnecessarily replace perfectly safe pipe. These funds could be better used to address other aspects of our safety systems.

INGAA also estimates that class change pipe replacements under the current regulations result in up to 800 million standard cubic feet of natural gas blowdowns to the atmosphere annually. To put that into perspective, this quantity of gas could meet the needs of over 10,000 homes for a year and has the same GHG reduction benefits of removing 80,000 cars from the road. The single best way to further reduce methane emissions by the natural gas pipeline industry is to decrease the number of "blow downs" or voluntary releases of gas. Finalizing the rulemaking would substantially decrease methane emissions by stopping these unnecessary releases of gas into the atmosphere.

Historically, in place of a class location pipeline replacement change, INGAA members have submitted special permit applications to prove the safety of their pipes. However, these applications are burdensome to not only the industry, but also to PHMSA. Some of the problems include the process changing regularly and that it can take up to three years to approve a single permit. Finalizing this rule will provide regulatory certainty and consistency for industry stakeholders and the regulator.

INGAA is hopeful that PHMSA will hold a GPAC meeting as soon as possible and issue this crucial rule to improve safety and meet the collective goal of industry and the Biden Administration to lower GHG emissions.

3. The Gas Pipeline Advisory Committee (GPAC) Strengthens Rulemakings and Should Meet More Frequently

The GPAC is an advisory committee to the Department of Transportation and PHMSA on matters of natural gas pipeline safety and regulatory oversight. The GPAC is comprised of 15 members, with equal representation from the natural gas industry, federal and state agencies, and the public (such as safety advocates and emergency managers). GPAC's stated role is to review PHMSA's proposed regulatory initiatives to ensure the technical feasibility, reasonableness, cost-effectiveness, and practicability of each proposal. PHMSA is not bound by GPAC recommendations but must include its rationale for disagreeing with the recommendations in the preamble text of final rules. These processes are required by statute. GPAC can play an important role in completing our collective objective to enhance gas pipeline safety regulations. The time needed to complete a rulemaking is affected, in part, by the quantity and quality of dialogue with impacted stakeholders. Their dialogue is especially important when the subject of a rulemaking is a complex, technical topic such as pipeline safety regulation. New rules should leverage stakeholder knowledge and expertise to facilitate the deployment of new technologies and practices that are more effective and efficient, and less disruptive than the legacy methods that may be reflected in existing regulations.

Until recently, GPAC met regularly to consider important rules and discuss important safety advancements. However, since January 2021, the GPAC has only met once. With the known benefits of GPAC, INGAA believes that Congress should consider requiring PHMSA to hold at least two GPAC meetings per year.

Furthermore, PHMSA has chosen to disagree with a number of unanimous GPAC recommendations to multiple recent important final rules. While INGAA does not challenge PHMSA's independence to make these decisions, we believe that Congress can strengthen transparency by receiving detailed briefings from PHMSA on their rationales for these conclusions soon after issuing final rules.

4. PHMSA Should Prioritize Hiring Safety Engineers to Assist with Its Rulemakings

As I stated earlier, INGAA appreciates the important role of having a strong safety regulator. An effective way to remain strong is for PHMSA to invest in hiring qualified safety engineers to assist with writing rules.

INGAA recognizes that pipeline safety regulations are complicated and often take years of work to be drafted and implemented. To help assist with the rulemaking backlogs that frequently occur, adding safety engineers to PHMSA's team who already understand pipeline systems and federal code would be a great asset to the agency.

INGAA was pleased that Congress included in Section 102 of the PIPES Act of 2020 provisions requiring PHMSA to hire "eight full-time employees with subject matter expertise in pipeline safety, pipeline facilities, and pipeline systems to finalize outstanding rulemakings and fulfill congressional mandates." We are hopeful that PHMSA can successfully hire safety engineers to fulfil this requirement in a timely manner.

In conclusion, I truly appreciate the opportunity to testify in front of your Subcommittee today. Your efforts are critical to ensuring that PHMSA has the resources and direction necessary to continually improve safety in our industry.

Mr. NEHLS. Thank you, Mr. Grubb.

Mr. Caram, you are recognized for 5 minutes.

Testimony of Bill Caram, Executive Director, Pipeline Safety Trust

Mr. CARAM. Thank you, Chairman Nehls, Ranking Member Payne, and members of the subcommittee for the opportunity to speak today on pipeline safety.

My organization, the Pipeline Safety Trust, was formed after the devastating Olympic pipeline tragedy that stole the lives of three boys in Bellingham, Washington, in 1999, a tragedy Ranking Member Larsen so eloquently described.

The U.S. Justice Department was so appalled at the operations of the pipeline company and the lax oversight from the Federal Government that they asked the Federal courts to set aside money from the settlement to create the Pipeline Safety Trust as an independent national watchdog over the pipeline industry and its regulators. And it is with that spirit that I am speaking to you today. Everyone here today values safe communities, and we all share the goal of zero incidents. However, pipeline failures continue unabated, and we now face a potential new generation of nonhydrocarbon pipelines posing their own unique risks to our communities, and making our goal of zero incidents even further out of reach.

Since the last pipeline safety legislation was passed in December of 2020, a bit over 2 years ago, our Nation's pipelines have had over 1,300 reportable failures, more than 1 per day, killed or injured to the point of inpatient hospitalization 74 people, and caused nearly $1 billion in property damage. These failures include recent tragedies like the El Paso gas transmission pipeline explosion that killed two, including a 14-year-old girl, and the Consumers Energy explosion in Flint, Michigan, that killed two, including a 3-year-old girl, and also many high-profile hazardous liquid failures spilling millions of gallons of oil and gasoline into our rivers, oceans, and groundwater, including Colonial Pipeline's more than 2-million-gallon gasoline spill in a nature preserve in Huntersville, North Carolina.

To give a brief overview of our Nation's state of pipeline safety, I am sad to say that, when looking at the data, we haven't made much progress. We can, of course, pick out time periods to show trends either going up or down. However, going back to 2010, which I believe to be an objective starting point, since that is when PHMSA changed some of its reporting criteria, incident trends on all categories of pipelines, including incidents impacting people or the environment, are statistically flat. We are not making progress. To start making a real difference on pipeline safety. Congress needs to remove the statutory handcuffs it has placed on the rules PHMSA can produce. PHMSA is the only agency with a statutory cost-benefit requirement, an especially difficult hurdle to clear for infrastructure with low probabilities of failure, yet enormously devastating consequences.

PHMSA is also prohibited by statute from adopting design, initial testing, and construction standards that would apply to the existing 3 million miles of existing pipelines. Because of this non- application clause, PHMSA has been unable to require rupture mitigation valves in high-consequence areas—an NTSB recommendation after PG&E's pipeline tragedy in San Bruno, California, in 2010. Of course, it is the aging infrastructure of existing pipelines that need this kind of lifesaving technology the most.

I want to take a moment to discuss one specific pipeline failure from 2020, a carbon dioxide pipeline operated by Denbury that failed near Satartia, Mississippi. Carbon dioxide is an asphyxiant that is heavier than air, and can therefore travel in dangerous and even lethal concentrations for long distances, displacing oxygen.

Nearly the entire community of 200 in Satartia was evacuated, with 45 of them seeking treatment at the hospital. Residents lost consciousness, had seizures, foamed at the mouth. First responders heroically pulled many to safety while donning scuba gear, probably saving lives, their truck engines unable to operate in the oxygen-deprived environment. The community of Satartia is incredibly lucky to not have suffered any fatalities.

As Congress has highly incentivized carbon capture and sequestration, communities are looking at a potentially enormous buildout of CO2 pipelines. Currently relatively rare and remote, experts are predicting a future with 20 times or more the amount of CO2 pipelines, and much closer to communities.

On top of this, they are dangerously unregulated. Current PHMSA standards only regulate CO2 pipelines if in a supercritical fluid with more than

90 percent purity. Therefore, gas and liquid CO2 pipelines are unregulated. Many man-made sources of CO2 can have much higher levels of impurities that pose public health and pipeline integrity issues such as corrosion. There are no PHMSA standards on levels of these dangerous impurities. Congress should mandate that PHMSA regulate these and the other equally frightening regulatory gaps that need to be addressed before the public can have any kind of confidence in a potential national buildout of CO2 pipelines.

In order to honor the value we place on safe communities, Congress needs to give PHMSA more authority, more resources, and implement the safety recommendations I have detailed in my written testimony.

Thank you again for inviting me to testify today. I look forward to answering any questions the subcommittee may have and helping Congress improve the Nation's state of pipeline safety.

[Mr. Caram's prepared statement follows:]

Prepared Statement of Bill Caram, Executive Director, Pipeline Safety Trust

Good morning, Chairman Nehls, Ranking Member Payne, Committee Chair Graves, Committee Ranking Member Larsen, and members of the Committee. Thank you for inviting me to speak today on the vital subject of pipeline safety. My name is Bill Caram, and I am the Executive Director of the Pipeline Safety Trust. The Pipeline Safety Trust was created after the Olympic Pipe Line tragedy in Bellingham, Washington in 1999. That entirely preventable failure spilled nearly a quarter-million gallons of gasoline into a beautiful salmon stream in the heart of our community which eventually ignited and killed three boys. The U.S. Justice Department was so appalled at the operations of the pipeline company and equally appalled at the lax oversight from the federal government, that they asked the federal courts to set aside money from the settlement to create the Pipeline Safety Trust as an independent national watchdog organization over the pipeline industry and its regulators.

We work to ensure that no other community must endure the senseless grief that Bellingham has had to experience from a pipeline tragedy. Sadly, there have been many senseless pipeline tragedies and disasters since Bellingham. I am here today, hoping that we can continue to work together in a bipartisan way, to help us move towards our shared goal of zero incidents. Today I would like to focus my testimony on:

- Overview of the state of U.S. pipeline safety
- Critical pipeline safety issues
    - Eliminate cost-benefit requirements under 49 U.S.C. § 60102
    - Eliminate the nonapplication clause in 49 U.S.C. § 60104(b)
    - Include mandamus clause
    - Prohibit reportable unintended releases
    - Increase authorized appropriations and add recruitment and retention flexibility
    - Require rupture mitigation valves on existing gas pipelines in High Consequence Areas
    - Improve carbon dioxide pipeline safety regulations
    - Improve hydrogen pipeline safety
    - Improve geohazard mitigation regulations
    - Natural gas incident reporting
- Public transparency improvements
    - National Pipeline Mapping System (NPMS) Improvements
    - Require operators to disclose certain safety information
    - Improve reporting data metrics
    - Create Office of Public Engagement
- Other needed safety improvements
    - Increase penalties
    - Eliminate natural gas operator's choice in determining High Consequence Areas
    - Close class location loophole on building occupancy
    - Eliminate safety related condition report exemptions
    - Require mandatory reporting of liquid over-pressurization events
    - Require improvements to state 811 damage prevention programs
    - Mandate offshore pipeline safety improvements
    - Clarify "confirmed discovery" definition
- Appendix
    - Statutory and regulatory language where appropriate

Overview of the State of U.S. Pipeline Safety

Since Congress passed the PIPES Act of 2020, a little over two years ago, there have been 1,300 reportable pipeline failures, more than one per day, 74 people have been either killed or injured to the point of inpatient hospitalization, and nearly $1 Billion in property damage.

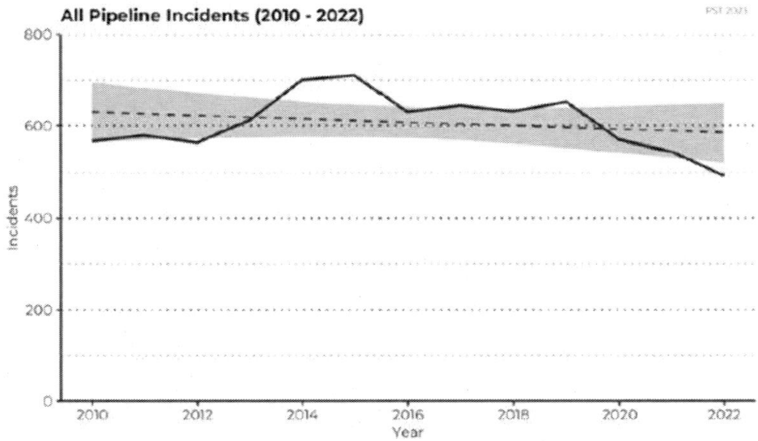

Source: PHMSA Incident and Mileage Data (2023).

While everyone on today's panel supports the goal of zero incidents, unfortunately, we have a long way to go. While you can slice and dice data opportunistically to demonstrate progress, when you look at the PHMSA reported data objectively, we are not making real progress on pipeline safety. My organization looked at the data going back to 2010 since that is when PHMSA changed some reporting. That is the longest period we can analyze without some data manipulation, and we believe that to be an objective starting point. Total incidents for gas and hazardous liquids show a trend line going down very slightly—a basically flat line with no real progress over the past twelve years.

Filtering for only those incidents deemed "significant" by PHMSA, we see a trend that is slightly increasing. For all the progress the industry touts on technological advancements and safety management systems, we are not moving towards our target of zero incidents.

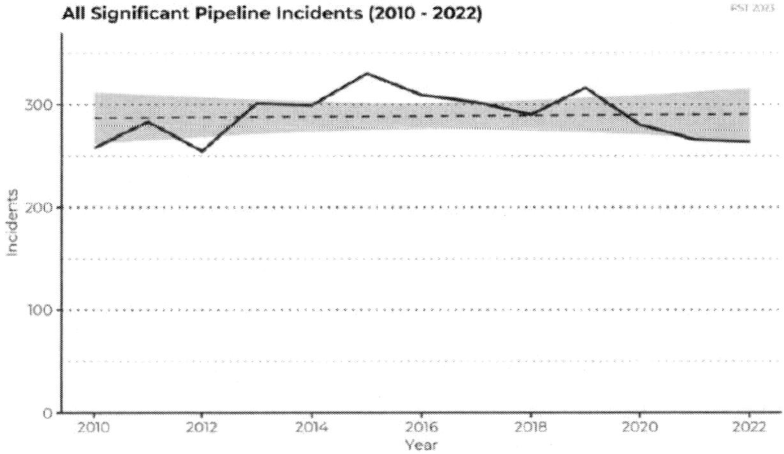

Source: PHMSA Incident and Mileage Data (2023).

Also of concern is the fact that approximately two-thirds of all incidents and significant incidents are from causes that are under the operator's direct control such as corrosion, incorrect operations, equipment failures, and problems with materials, welds, and equipment.

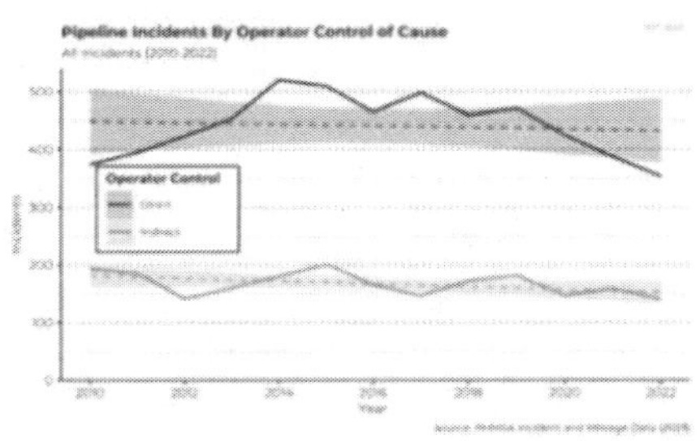

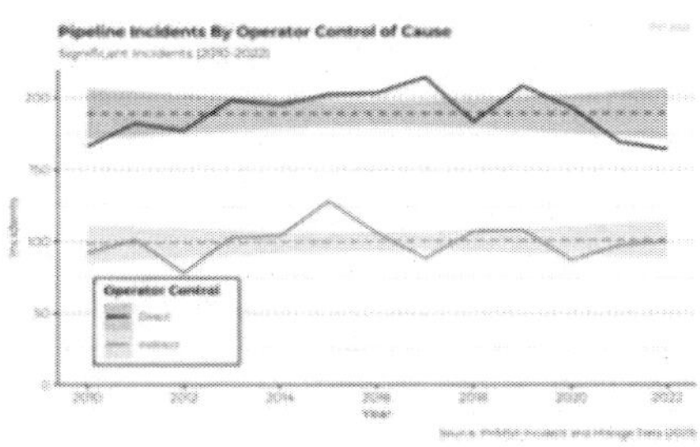

Over the past twenty years, regulators and industry have focused much emphasis in reducing pipeline incidents on "Integrity Management" efforts in "High Consequence Areas." The theory behind Integrity Management programs makes perfect sense—focus efforts in those areas where the most harm to people and the environment may occur, work hard to identify all risks in those areas, put into place programs to test for and mitigate those risks, and implement a continuous improvement program to drive down the number of failures.

Unfortunately, for both hazardous liquid and gas transmission pipelines these Integrity Management programs do not seem to have lived up to their promise. Incident rates within High Consequence Areas as compared to outside HCAs continue to climb in the case of hazardous liquid pipelines and do no better with regards to gas transmission pipelines. These two graphs, generated from PHMSA's Integrity Management Data, demonstrate our concern with current integrity management programs. Some in the industry argue that older, prescriptive class location rules can now be relaxed because of the implementation of Integrity Management, but as the graphs show: It is too early to go to a performance-based Integrity Management system until the industry can prove that Integrity Management works as it should.

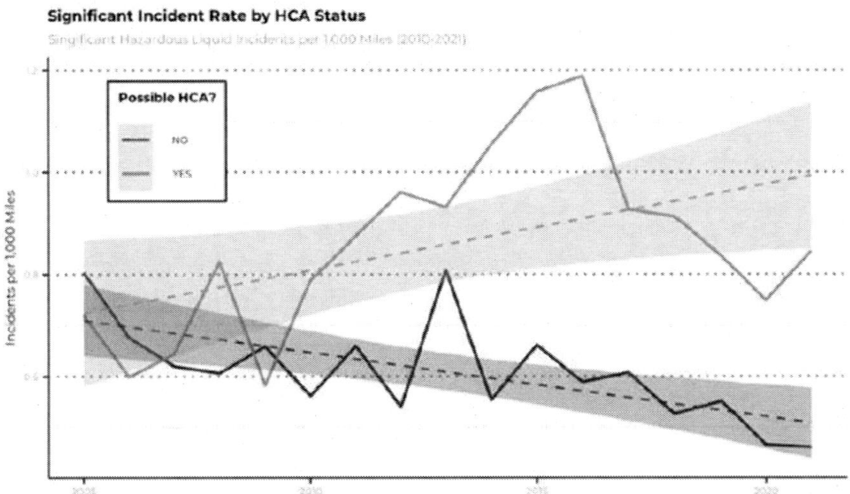

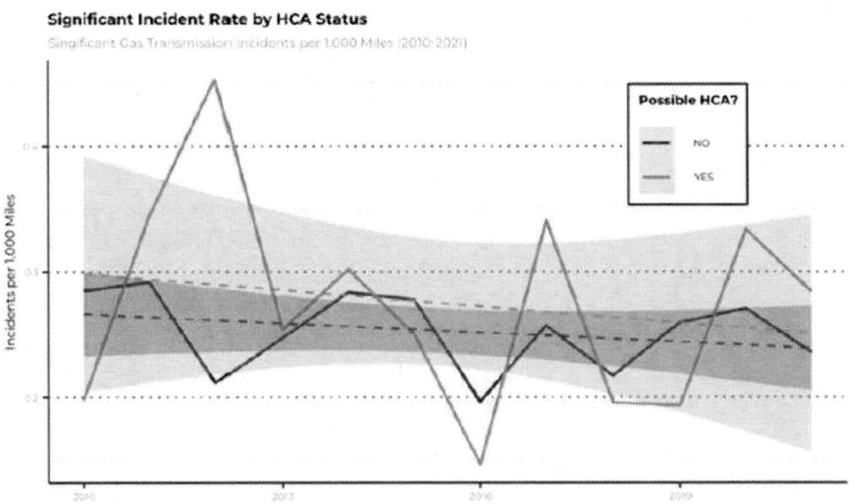

The below chart visualizes the ratio of incident rates inside HCAs vs outside. Values above zero mean that HCA rates are worse inside an HCA vs outside, meaning Integrity Management programs are not working.

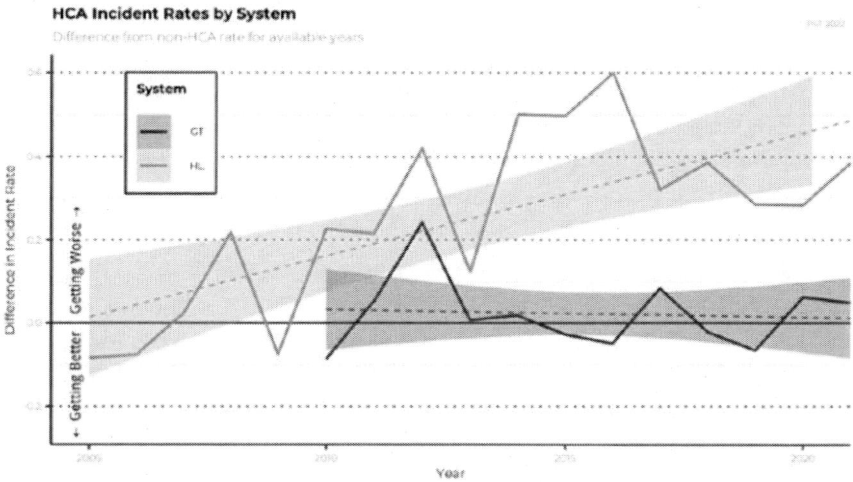

Incident Rate = Incidents per 1,000 Miles.
Source: PHMSA Annual Reports, Flagged Files, and Mileage Data (2005–Present).

Critical Pipeline Safety Issues

Please note, suggested statutory and regulatory language is provided for each issue, when applicable, in the appendix at the end of this testimony.

Eliminate Cost-Benefit Requirements Under 49 U.S.C. § 60102

PHMSA rulemaking is subject to two sets of cost-benefit requirements: one under the Pipeline Safety Act and one under Executive Order 12866, which requires an economic analysis of every major rule reviewed by the Office of Management and Budget. While the additional analysis does not mandate that the benefit of new regulations outweigh the cost, that is often how the industry and PHMSA itself views this requirement—making passage of new regulations difficult or nearly impossible in some areas. In fact, the industry, represented by American Petroleum Institute (API) and GPA Midstream, are suing PHMSA over its new gas gathering rule.[79]

[79] Tom DiChristopher, Pipeline Industry Takes Dispute Over US Gathering Line Rule to Court, S&P GLOBAL (June 7, 2022) https://www.spglobal.com/marketintelligence/en/news-

In 1996, a concerted Congressional effort was made to insert cost-benefit analysis requirements into rulemaking requirements under a whole host of environmental protection and health statutes, presumably to reduce regulatory burden and codify the requirements for regulatory cost benefit analyses put in place by Presidents Reagan and Clinton in Executive Orders. Those Congressional efforts ultimately fell short of widespread success because so many members of Congress realized how such measures in the statute would provide a well-funded industry a strong litigation hook that would make easy to challenge new regulations and nearly impossible to protect people's health and safety. The 1996 reauthorization of the pipeline safety program, based solely on timing, represents the only health and safety or environmental protection statute where such an explicit directive to an administrative agency to base regulation of risk on a cost-benefit test was inserted into law.[80]

We urge Congress to put PHMSA's rulemaking on an even playing field with all other agencies by amending 49 U.S.C. § 60102 to eliminate references to the risk assessment/cost-benefit analysis in § 60102(b)(2)(D) and (E); § 60102(b)(3), (4), (5) and (6). PHMSA would remain subject to the requirements of the Executive Orders requiring a cost benefit analysis of major rules proposed by any agency, and the requirements for transparency in rulemaking provided by the existing statute and procedures.

Eliminate the Nonapplication Clause in 49 U.S.C. § 60104(b)

49 U.S.C. § 60104(b) specifically prohibits PHMSA from adopting a design, installation, construction, initial inspection, or initial testing standard from applying to existing pipelines.

After PGE's tragic failure in San Bruno, CA, when operators were unable to close valves and isolate the fuel feeding the blowtorch destroying a neighborhood for nearly two hours, NTSB recommended PHMSA require operators to install Automatic- Shut-Off or Remote-Controlled Valves in all High-Consequence Areas (HCAs), including existing pipelines.[81] Even if such a regulation could survive the statutory cost-benefit requirement, it would be prohibited by section 60104(b). This means that despite the fact that the science behind safe pipeline operation continues to develop, there will almost

insights/latest-news-headlines/pipeline-industry-takes-dispute-over-us-gathering-line-rule-to-court- 70713022.

[80] Sara Gosman, Justifying Safety: The Paradox of Rationality, SOCIAL SCI. RES. NETWORK (Apr. 22, 2017).

[81] Nat'l Transp. Safety Bd., Accident Report: Pacific Gas and Electric Company Natural Gas Transmission Pipeline Rupture and Fire San Bruno, California September 9, 2010, NSTB/PAR– 11/01 (Aug. 30, 2011) https://tinyurl.com/56tfuw9w.

always be thousands or even millions of miles of operational pipelines to which improved safety standards will never apply. Often, it is the ageing pipelines that need these minimum safety improvements the most. Additionally, this is a critical problem at this moment in history, when congressional investments have been made that have spurred interest in developing carbon dioxide and hydrogen pipelines. Because of the nonapplication clause, if PHMSA does modernize its woefully out-of-date CO2 pipeline construction standards or develop special standards for hydrogen pipelines prior to their construction, the regulations will not apply to any pipelines already in the ground.

Congress should eliminate the nonapplication clause found at 49 U.S.C. §60104(b) to ensure that design, installation, construction, initial inspection, and initial testing standards can apply to existing pipelines when appropriate.

Include Mandamus Clause

In 2015, the City of San Francisco, after witnessing the terrible nearby tragedy in San Bruno, felt so strongly that PHMSA was failing to uphold the statutory requirements and Congressional mandates under the Pipeline Safety Act that it went to court to force PHMSA to do so. The Ninth Circuit Court of Appeals, without addressing the merits of the case, dismissed the case with an opinion holding that the Pipeline Safety Act does not provide the basis of a mandamus action to force PHMSA to carry out a duty under the Act.[82] The court relied, in part, on the absence of any explicit mandamus remedy at 49 U.S.C. § 60121 ("Actions by private persons").

The Trust strongly believes that local and state governments, and others, should be able to ask the courts to carry out what Congress has required of it in statute. This is a common protection in many other laws. We urge Congress to include the following language in this year's reauthorization to close this loophole.

Prohibit Reportable Unintended Releases

In 2013, a major failure occurred on ExxonMobil's Pegasus Pipeline in Arkansas causing 134,000 gallons of crude oil to spill into a neighborhood, contaminating homes and yards, a creek, wetlands, and Lake Conway. In a review of the PHMSA enforcement action following the 2013 spill, the Fifth Circuit found that an operator can cause a reportable incident, or even a

[82] City & County of San Francisco v. U.S. Dep't of Transp., No. 12-cv-0711 (N.D. Cal. Feb. 28, 2013) (granting motion to dismiss) https://tinyurl.com/kecae69f.

significant incident, without necessarily having violated a safety regulation.[83] As written, the pipeline safety statutes do not expressly prohibit the release of gas or hazardous liquid from a pipeline.

To close that loophole, the Pipeline Safety Trust proposes that section 60118 be amended to require operators to avoid releases of gas or hazardous liquids in quantities that would make them reportable incidents under PHMSA regulations. This section is subject to enforcement by PHMSA under § 60122 or by the Attorney General under § 60120.

Increase Authorized Appropriations and Add Recruitment and Retention Flexibility

PHMSA, already a notoriously underfunded and understaffed agency, has had large increases in Congressional mandates without a corresponding increase in funding. For example, nearly 100,000 miles of gas gathering lines have finally come under PHMSA regulations and another approximately 300,000 miles are under new reporting requirements. Also on the horizon is a new generation of pipelines carrying carbon dioxide and hydrogen, requiring new expertise and personnel. State programs, responsible for oversight of more than 80% of the nation's pipeline mileage, are also feeling the squeeze on their capacity.

PHMSA has long been considered underfunded and understaffed and therefore reliant on the industry it is tasked to regulate for technical expertise on rulemaking. A 2015 Politico investigation[84] found that PHMSA is an agency "that lacks the man-power to inspect the nation's. oil and gas lines, that grants the industry it regulates significant power to influence the rule-making process, and that has stubbornly failed to take a more aggressive regulatory role, even when ordered by Congress to do so." PHMSA has also long had difficulty in attracting and retaining experienced personnel as the industry often hires staff away at higher salaries.

Critical components to changing this culture are authorizing significantly more funding and allowing more flexibility in the recruitment and retainment of experienced personnel. We also recommend a significant increase to authorized fundingof PHMSA's state programs.

---

[83] ExxonMobil Pipeline Co. v. U.S. Dep't of Transp., Order on Petition for Review, No. 16–60448 (Aug. 14, 2017) https://www.ca5.uscourts.gov/opinions/pub/16/16-60448-CV0. pdf.

[84] Andrea Restuccia and Elana Shor, *Pipelines Blow Up and People Die*, POLITICO (Apr. 21, 2015) https://www.politico.com/story/2015/04/the-little-pipeline-agency-that-couldnt-217227.

Congress should, when amending Section 60125 of title 49, subsection (a), includea substantial increase to PHMSA's authorized funding to reflect the enormous increase in their charge as previously described. Congress should also include a substantial increase for the State Pipeline Safety Grant Program authorized in Section 60107 of title 49.

Require Rupture Mitigation Valves on Existing Gas Pipelines in High Consequence Areas

Advancements to rupture mitigation valve technology have been made and adopted into PHMSA's regulations, but these regulations do not apply to existing pipelines, even on older pipes in areas that could affect densely populated areas. Arguably these are the pipelines that need this technology the most.

In 2022, PHMSA revised its pipeline safety regulations to require rupture mitigation valves (RMVs), or alternative equivalent technologies, to newly constructed or entirely replaced onshore gas transmission, Type A gas gathering, and hazardous liquid pipelines with diameters of 6 inches or greater.[85] The rule did not, however, require operators to retrofit older pipes because of the nonapplication clause found at 49 U.S.C. §'104(b), which prohibits the Pipeline and Hazardous Materials Safety Administration (PHMSA) from promulgating regulations to existing facilities. Because of this, PHMSA fell short of adequately implementing NTSB's recommendation.[86] Excluding certain pipelines from implementation of critical safety technology based on age is dangerous. Older pipes are likely more prone to failure, and it is arbitrary to require critical safety technology on some but not all pipelines. Requiring operator to retrofit older pipelines with RMVs in HCAs would protect areas with more people and buildings that could be affected by a failure. 49 C.F.R. § 192.903. Because of the nonapplication clause, however, Congress must draft self-executing language for PHMSA to have the authority to promulgate these regulations. Suggested language is provided in the appendix.

---

[85] Pipeline Safety: Requirement of Valve Installation and Minimum Rupture Detection Stand ards, 87 Fed. Reg. 20,940–992 (Apr. 8, 2022).

[86] Nat'l Transp. Safety Bd., *Press Release: NTSB Issues Response to PHMSA's Valve and Rupture Detection Rule*, (Apr. 1, 2022) https://www.ntsb.gov/news/press-releases/Pages/NR20220401B.aspx.

Improve Carbon Dioxide Pipeline Safety Regulations

Given the Congressional incentives driving carbon capture and sequestration investment, many experts expect a large increase in the mileage of the nation's carbon dioxide (CO2) pipelines. Once relatively rare and remote, these pipelines will soon be much closer to people and communities. The Denbury CO2 pipeline failure in Satartia, MS demonstrated the unique safety risks that these pipelines pose. An asphyxiant that is heavier than air, CO2 can move as a plume in a dangerous and even lethal concentration close to the ground for long distances after a failure. Current PHMSA safety regulations are inappropriate and insufficient, as described in a Pipeline Safety Trust report.[87]

The current definition of "carbon dioxide" in the federal pipeline safety regulation does not apply to all CO2 pipelines that may be developed for CCS projects.

Currently, only CO2 that is moved in a supercritical state is regulated under the current definition, meaning gaseous and liquid CO2 pipelines are not currently regulated.

There is currently no defined safe distance or plume dispersion model for developing a potential impact radius (PIR) along CO2 pipelines.

CO2 has unique physical properties which warrant the development of a unique PIR zone to be promulgated into federal pipeline regulation.

There is currently no requirement to add an odorant to transported CO2.

Carbon dioxide is odorless, colorless, doesn't burn, and is heavier than air meaning that releases are harder to observe and therefore avoid.

The unique physical properties of CO2 moved at high pressures through pipelines can cause running ductile fractures upon rupturing.

This essentially means that a pipe has a higher likelihood of opening up like a zipper when a rupture occurs, leading to more product being released over a shorter period of time and potentially violent and dangerous pipe shrapnel.

Contaminants within CO2 products being transported can jeopardize the integrity of the pipeline.

Water, when mixed with carbon dioxide, can form carbonic acid which can rapidly erode carbon steel.

Different industries can produce numerous other contaminants, including SOx and NOx, which can be toxic to public health, affect the temperature and

[87] Accufacts, Inc., *Accufacts' Perspectives on the State of Federal Carbon Dioxide Transmission Pipeline Safety Regulations as it Relates to Carbon Capture, Utilization, and Sequestration within the U.S.* (Mar. 23, 2022) https://pstrust.org/wp-content/uploads/2022/03/3-23-22-Final-Accufacts-CO2-Pipeline-Report2.pdf.

pressure of the product, and/or cause corrosion, potentially impacting the safe operation of the pipeline.

The risks associated with the conversion of existing transmission pipelines to CO2 service have not been fully investigated.

Given the unique properties of CO2 mentioned previously, pipeline conversions have the potential to be at higher risk of failure from CO2 service than conventional hydrocarbon or even new construction CO2 pipelines.

For the public to have any confidence in the safety of these pipelines proposed through communities, regulations need to be modernized. However, given the small number of existing mileage of CO2 pipelines, PHMSA may not have enough information to preemptively justify the cost of such improvements.

Congress should require PHMSA to promulgate rules addressing each of the above-listed regulatory gaps. Given CO2's physical properties, unique safety risks, and ability to be transported in multiple phases, PHMSA should allot CO2 its own section of code, CFR Part 197. These rules should not be subject to PHMSA's statutory cost-benefit requirement.

Improve Hydrogen Pipeline Safety

Hydrogen has been highly incentivized in recent legislation such as the Production Tax Credit in the Inflation Reduction Act. Gas distribution operators are considering blending hydrogen into existing gas distribution infrastructure and the trade group the American Gas Association includes hydrogen blends of 20% as a key component of their Net Zero plan for the industry.[88] However, hydrogen transportation by pipeline poses many safety risks and key knowledge gaps remain. The risks run highest when the pipelines are near people. At least one operator in Hawaii has blended hydrogen, however that system is unique enough that it likely cannot serve as a model for the rest of the country.

Hydrogen has a much higher flammability range than methane and is known to embrittle certain types of steel pipelines. A report on blending hydrogen commissioned by the California Public Utility Commission from University of California Riverside found an alarming number of safety risks and knowledge gaps. A report by Accufacts commissioned by the Pipeline

---

[88] American Gas Association, *Net Zero Emissions Opportunities for Gas Utilities* (Feb. 8, 2022) https://www.aga.org/wp-content/uploads/2022/02/aga-net-zero-emissions-oppor tunities-for-gas-utilities.pdf.

Safety Trust[89] stated that the weakest safety link for hydrogen blends in the distribution system were the pipes inside residences. Additionally, hydrogen has less energy density by volume of methane, so any blend will only deliver about a third of the greenhouse gas emissions (e.g., a 20% blend of hydrogen will reduce greenhouse gas emissions by less than 7%). Hydrogen is also a potent indirect greenhouse gas itself with a propensity to leak, therefore leaks could quickly erode all the intended climate benefits.[90]

Congress should prohibit new hydrogen blends in gas distribution systems until the National Academy of Sciences has issued a report from both a safety and climate perspective.

Require Blended Products to be Reported to PHMSA

An operator is only required to report the "predominant product" in a natural gas pipeline system to PHMSA. This has been interpreted to mean only reporting a product that is > 50% present, overwhelmingly methane/natural gas.

Currently operators blend products such as propane or hydrogen into existing systems at unknown rates. In December 2022, CenterPoint Energy blended propane into its Southern Indiana natural gas distribution system incorrectly and triggered hundreds of carbon monoxide events, sending four people to the hospital.[91] One operator in Hawaii is blending hydrogen into its gas distribution system, which we only known because they have volunteered the information.

Congress should require operators to report to PHMSA blended, non-predominant products that at any point in time exceed 3% by volume.

Improve Geohazard Mitigation Regulations

There have been a number of recent, serious pipeline failures due to land movement and other geological hazards. The 2020 Enbridge failure in Hillsboro, Kentucky; the 2020 Denbury CO2 pipeline failure in Satartia, MS (45 people sought treatment at the hospital); and the 2022 Marathon Pipe Line spill in Edwardsville, IL (165,000 gallons of crude spilled in and near creek)

---

[89] Accufacts, Inc., *Safety of Hydrogen Transportations by Gas Pipeline* (Nov. 28, 2022) https://pstrust.org/wp-content/uploads/2022/ 11/11-28-22-Final-Accufacts-Hydrogen-Pipeline-Report.pdf.

[90] Pipeline Safety Trust, *Hydrogen Pipeline Safety Summary for Policymakers* https://pstrust.org/wp-content/uploads/2023/01/hydrogenlpipelinelsafetylsummaryl1118l23.pdf.

[91] Pipeline Safety Trust, CenterPoint Energy's Apology Not Enough (Feb. 8, 2023) https://pstrust.org/centerpoint-energys-apology-not-enough-more-must-be-done-to-protect-our-commu- nities-from-pipeline-incidents/

were all due to land movement. PHMSA has issued multiple Advisory Bulletins to operators on geohazard threat mitigation. Operators are required to mitigate against any threat within High Consequence Areas, but do not have any specific requirement to mitigate against geohazards outside of those areas. If we are committed to zero incidents, we need to address the risk of geohazards such as land movement, river scouring, and other geologic threats to pipeline integrity.

Congress should amend 49 U.S.C. § 60108 to require operators to include geohazard mitigation in their inspection and maintenance plans.

Natural Gas Incident Reporting

PHMSA can only regulate against issues that it is aware of. Unfortunately, short- comings in PHMSA's incident reporting regulations keep it in the dark because its regulations only require reporting if certain thresholds are met. Consequently, many large and potentially dangerous incidents are not reported to the administration. This means that that PHMSA's safety data likely underrepresents incident prevalence and that the opportunity to use these incidents as a learning opportunity is lost. The Pipeline Safety Trust recommends that Congress direct PHMSA to amend part 191 of its pipeline regulations and reporting forms to modernize the requirements for reportable incidents. More detail, statutory language, and proposed regulatory amendments are provided in the appendix.

Reporting of Fires and Explosions—Gas pipeline leaks are far more likely to result in immediate combustion and fire than hazardous liquid leaks. This places public safety and the environment at risk. Yet unlike hazardous liquid pipeline operators, gas pipeline operators are not required to report incidents which result in fire or explosion that do not meet other reporting requirements. 49 C.F.R. § 191.5; § 191.3. Congress should require PHMSA to make reporting of fires and explosions associated with gas pipelines mandatory.

Property Damage Thresholds—Until recently, the property damage thresholds for reporting incidents to PHMSA was $50,000 for both gas and hazardous liquid pipelines. However, in 2021, PHMSA issued final rule in response to industry feedback that the threshold was too low for gas pipelines. 86 Fed. Reg. 2219. This rule increased the gas incident reporting threshold for property damage to $122,000, to be adjusted annually for inflation. 49 C.F.R. pt. 191, app'x. With record inflation, the current threshold stands at a staggering $129,300. The gas rule excludes the value of the gas itself, which is also distinct from the liquid rule.

There is no reason the property damage incident reporting thresholds to differ to such an extreme. This is especially true given the fact that methane is a major contributor to climate change and presents a dangerous threat to the public when leaked from pipeline infrastructure. Congress should require PHMSA to make the threshold for reporting of property damage for hazardous liquid and natural gas pipelines equal by lowering the property damage threshold for natural gas incidents back to $50,000 and require that the cost of lost product be included in this calculation. $50,000 is still a substantial amount of money for a member of the public, even if it is not for wealthy oil and gas companies.

Reporting of Gas Releases—PHMSA regulations require hazardous liquid releases as small as 5 gallons to be reported. 49 C.F.R. § 195.50(b). By comparison, natural gas regulations, drafted before the collective consensus that methane emissions are a major contributor to climate change, are extremely permissible, requiring reporting only if an incident is an "unintentional estimated" release of three million cubic feet or more. Id. at § 191.3(1)(iii). Not only is the release required to be unintentional, but the threshold is unjustifiably high. For context, an operator can release enough gas to power over 17,000 U.S. homes without reporting the incident to PHMSA.[92]

Congress should require PHMSA to acknowledge the seriousness of methane emissions and reduce the reporting threshold for gas pipelines to 50,000 cubic feet, regardless of intent. Operators are already required to minimize intentional and accidental releases,[93] they should already have capacity to monitor for releases and be required to report them to PHMSA.

Public Transparency Improvements

National Pipeline Mapping System (NPMS) Improvements

PHMSA's National Pipeline Mapping System (NPMS) is one of the main ways the public can learn about pipelines in their area. However, they are often left in the dark with much needed information hidden from public view.

NPMS is a dataset containing locations of and information about gas transmission and hazardous liquid pipelines and Liquefied Natural Gas (LNG) plants which are under the jurisdiction of PHMSA. The NPMS also contains

[92] Am. Gas Ass'n, Natural Gas: The Facts (2019) https://tinyurl.com/sfhm36nv ("On a daily basis, the average U.S. home uses 175 cubic feet of natural gas.").

[93] Protecting our Infrastructure of Pipelines and Enhancing Safety Act of 2020, S. 2299, 116th Cong. § 114 (2020).

voluntarily submitted breakout tank data. The data is used by PHMSA for emergency response, pipeline inspections, regulatory management and compliance, and analysis purposes. It is used by government officials, pipeline operators, and the general public for a variety of tasks including emergency response, smart growth planning, critical infrastructure protection, and environmental protection.

NPMS offers operator and pipeline specific data to the public, first responders, and local governments. There are different versions of NPMS, depending on the user. The general public can see pipeline maps of one county at a time, with limited information about included pipelines and incidents. Approved government officials or pipeline operators gain access to more detailed pipeline maps with High Consequence Areas identified and additional scope and detail. Gathering pipelines and distribution pipelines are not included.

Both Congress, through statutory mandates, and the NTSB, through recommendations, have stressed the importance of public access to this information. The PST believes strongly in the supportive role the public can play as a partner in safer pipelines, but that partnership is only as good as the information the public can access. Given that, there are several shortfalls with NPMS. The current accuracy and detail of the NPMS data are not sufficient to adequately assist local communities who are planning or preparing for potential emergencies. Also, no HCAs are viewable on the public maps which is problematic and needs to be changed. In fact, there is already a statutory requirement, from the Pipeline Safety Regulatory Certainty and Job Creation Act of 2011, to incorporate HCAs into NPMS and update biennially. Congress should finish what it started and give the public, first responders, and local governments' access to this critical information.

Congress should also require the mapping of gathering pipelines. Gas gathering pipelines have grown in diameter and pressure in recent years and their safety risks can be indistinguishable from gas transmission pipelines in some cases. All users of NPMS need to be able to see where gathering lines are located.

Require Operators to Disclose Certain Safety Information

The public deserves more transparency about the levels of risk they face from pipelines in their communities and near their homes. Unfortunately, this information is often shielded from the public eye: Concerned citizens cannot obtain information about High Consequence Areas (HCAs), Medium Consequence Areas (MCAs), Potential Impact Radii (PIRs), or class locations nor can they obtain pipe size or pressure information, such as Maximum Operating Pressure (MOP) or Maximum Allowable Operating Pressure (MAOP). Operators are already submitting some of this information to PHMSA, but it only discloses minimal information to the public, such as approximate location and operator name via the NPMS. Allowing the public access to this information would significantly increase awareness regarding where integrity management is being implemented and allow them to weigh risks when making decisions such as where to live.

Congress should amend 49 U.S.C. § 601[94] to require operators to disclose pipeline safety and attribute information to those who inquire.

Improve Reporting Data Metrics

PHMSA can improve public engagement around pipelines by making the data available on its website easier for the public to digest and draw conclusions. Multi-stakeholder groups including the public, regulators, and industry met in 2015 and 2017 to develop performance measures for natural gas and hazardous liquid pipelines. The group working on hazardous liquid measures created the helpful metric of Accidents Impacting People or the Environment (IPEs). Performance measures for Highly Volatile Liquids (HVLs) or Liquified Natural Gas (LNG) have not been developed, and the performance measures developed for hazardous liquid pipelines do not align with those created for natural gas pipelines. Over the past few years, the pipeline industry has been developing new standards for pipeline safety performance measures that do not align with those of PHMSA, potentially creating more confusion than clarity regarding performance.

Congress should mandate PHMSA to convene multiple stakeholder groups to revisit the measures previously developed to assess their usefulness and effectiveness as well as develop new measures for HVLs and LNG. Stakeholders should include, at a minimum, Tribal governments, Tribal members, safety advocates, environmental advocates, state and federal regulators, and industry.

Create Office of Public Engagement

Public understanding and engagement are critical aspects in ensuring pipeline safety throughout the country. Currently, PHMSA, and more specifically the Office of Pipeline Safety, has no independent division to ensure effective

---

[94] PHMSA, Summary of Cases Involving Civil Penalties: Civil Penalties Resolved (2002–2021) https://tinyurl.com/dvw837tc.

public engagement and education in the pipeline safety process. PHMSA does have "Community Liaison Services" which are intended to help members of the public when contacted with questions related to pipeline safety, however, duc to lack of independence, training, and support from PHMSA, these services are significantly lacking in their ability to provide meaningful assistance to the public.

For members of the public to better understand and engage in the regulatory and safety aspects of pipeline awareness, Congress should direct PHMSA to create and fund an Office of Public Engagement. This independently run office would build on and enhance the effort of the already established PHMSA Community Liaison Services program by providing much needed support and two-way engagement directive for the administration.

The Office of Public Engagement could dispatch to communities after a pipeline failure to offer information and listen to residents' concerns. For a timely example, such an office could hold workshops across the Midwest and Gulf States to help educate members of the public on carbon dioxide pipelines and listen to the communities. Effective public engagement is vital to pipeline safety and an independent office dedicated to its values would help tremendously.

Other Needed Safety Improvements

Increase Penalties

PHMSA's penalty authority, and the agency's implementation of that authority, results in civil penalties that are economically insignificant to operators, are significantly smaller than those imposed by some states, and are disproportionate to the harm inflicted by pipeline failures. PHMSA's criminal penalty authority sets too high of a bar for criminal behavior and fails to hold companies accountable for criminal acts.

From 2002 to 2021, PHMSA's resolved civil penalty cases amounted to a mere $79,622,174—less than $4 million per year.16 By comparison, in the same period, 12,793 incidents have occurred killing 276 people, injuring 1,145, and causing over $10.1 billion dollars in property damage.[95] Despite PHMSA's 2017 maximum civil penalty adjustment to $209,002 for each day or $2,090,022 for a related series of violations, there has not been a significant

[95] PHMSA, Pipeline Incident 20 Year Trend (Jan. 23, 2023) https://tinyurl.com/5n6njd93.

increase in penalties proposed or collected, suggesting that PHMSA still remains reluctant to impose penalties In fact, some dramatic incidents have resulted in no civil penalties whatsoever. For example, just last year PHMSA imposed no penalties on operators responsible for a 165,000 gallon spill into an Illinois creek[96] and a methane release of approximately 1 billion cubic feet.[97]

Some states, notably California, have dramatically increased their use of civil penalties in the last decade, levying large fines like the one levied against PG&E following the San Bruno tragedy. The state regulator fined the utility $1.6 billion dollars for violations related to the 2010 failure in San Bruno and has since fined the utility additional millions relating to subsequent recordkeeping, reporting and other violations. These large fines are possible because the California and other state statutes do not have a limit on penalties for a related series of violations. Each day in violation is subject to another penalty.

Fortunately, it is very rare that a pipeline operator violates the regulations in a way that would be considered criminal. The Pipeline Safety Trust was born from one of those rare incidents where an operator's actions were proven to be so reckless as to kill members of the public and do uncounted environmental harm. The U.S. Department of Justice under President Bush did an outstanding job prosecuting that case, fining the company, and getting jail time for company employees.

There have only been a handful of other incidents caused by such reckless behavior from pipeline companies since that case nearly 20 years ago, but it is important to not create barriers that make it difficult to hold companies accountable when they knowingly or recklessly ignore the laws meant to keep people safe. The criminal statute applying to pipeline safety, 49 U.S.C. § 60123 requires that an operator "knowingly and willfully" violate the law—an unusually high bar for holding companies accountable for criminal behavior.

Congress should eliminate the cap on civil penalties for related series of violations and impose a mandatory minimum penalty for each violation. Congress should direct the Secretary to amend the agency's regulations

---

[96] NTSB, Marathon Pipe Line LLC Hazardous Liquids Pipeline Release, https://tinyurl.com/2d69wchm; PHMSA, Federal Enforcement Data, Marathon Pipe Line LLC (2006–2023) https:// tinyurl.com/4dansv3m (showing no penalties for 2022 pipeline incidents).

[97] 19 Letter from Robert Burrough, Director, Eastern Region Office of Pipeline Safety, PHMSA, to Clifford Baker, Senior Vice President, Equitrans Midstream Corporation (Dec. 29, 2022) https://tinyurl.com/2p9ekfck (proposing no fine).

accordingly within 180 days and align PHMSA's pipeline safety rules with its transportation of hazardous materials rules with respect to criminal penalties by amending section 60123 to adopt the "willfully or recklessly" language from 49 U.S.C. § 5124.

Eliminate Natural Gas Operator's Choice in Determining High Consequence Areas

Current federal regulations (49 C.F.R. § 192.905 and 192.903) allow for natural gas operators to choose between two methods in the identification of High Consequence Areas along the route of their pipeline.

High Consequence Areas are generally areas with higher populations in proximity to the pipeline. The chosen HCA method may be applied to the entire system, or different methods may be applied to different individual portions of the system. This discretion given operators not only creates inconsistency and uncertainty when PHMSA evaluates operator Integrity Management programs, but it also allows operators to choose whichever method requires the least effort and/or safety measures in their IM program.

The determination of a High Consequence Area should be limited to a singular definition. Specifically, by clarifying the definition of High Consequence Area in § 192.903.

Close Class Location Loophole on Building Occupancy

Under current regulations, gas transmission pipeline operators are required to classify their systems into Class Locations 1 through 4. These class locations generally signify how many buildings intended for human occupancy are located within the potential impact radius (PIR) of the pipeline and thus determine the level of safety requirements imposed on the operator for that section of pipeline. The regulation for determining class 3 areas creates a loophole which has the potential to exclude pipelines close to churches, theaters, and other public areas that may hold hundreds of people only a few days per week.

The class location of a gas transmission pipeline impacts the pressure at which the pipeline can operate and has other impacts on how an operator must comply with the PHMSA regulations. 49 C.F.R. § 192.5(b)(3)(ii) creates speculative criteria which limits the safety requirements associated with class 3 location areas. This section of the regulation partially defines a class 3 area as "An area where the pipeline lies within 100 yards (91 meters) of either a building or a small, well-defined outside area (such as a playground, recreation area, outdoor theater, or other place of public assembly) that is occupied by 20 or more persons on at least 5 days a week for 10 weeks in any 12-month period. (The days and weeks need not be consecutive.)" This regulation is not strict enough to ensure that pipelines that could endanger large numbers of people are held to higher safety standards.

Congress should require PHMSA to clarify and tighten this definition and regulation to close this loophole allowing operators to avoid stricter safety standards in areas with churches, playgrounds, and similar areas and buildings.

Eliminate Safety Related Condition Report Exemptions

Existing regulations requiring operators to disclose safety related conditions are ambiguous, lenient, and do not encompass all situations that warrant reporting.

49 U.S.C. § 60101(h) requires the Secretary to make rules requiring each operator of a pipeline facility to submit a written report to the Secretary on any (a) condition that is a hazard to life, property, or the environment; and (b) any safety related condition that causes or has caused a significant change or restriction in the operation of a pipeline facility. That written report must go to PHMSA and the state regulators within five days of the operator first establishing that the condition exists. However, the rules (49 C.F.R. § 191.23 and 195.55) enacted to implement that statute list only 8 specific kinds of safety related conditions, most with a large amount of operator discretion built into their definitions, and then provide a set of three reasons that even if the condition meets one of those eight requirements, a report isn't required.

Congress should require operators to submit reports to PHMSA on all safety-related conditions as originally mandated and make them easily available to the public.

Require Mandatory Reporting of Liquid Over-Pressurization Events

Require Improvements to State 811 Damage. Over-pressure events are a serious threat to pipeline safety that can adversely impact pipeline integrity and cause incidents that harm people and the environment. On June 10, 1999, an over-pressurization event occurred that changed the lives of many occurred in Bellingham, WA when the Olympic Pipe Line ruptured. The rupture leaked 277,200 gallons of gasoline into Hanna and Whatcom Creek, which flow through downtown Bellingham and directly into Bellingham Bay. The gasoline vapor subsequently ignited and exploded, killing three young men: Liam Wood, Wade King, and Stephen Tsiorvas. The cause of the incident was a failed pressure relief valve that caused the massive pressure surge and rupture.

To this day, despite the potential for disaster, operators of liquid pipelines are not required to report over-pressurization events to PHMSA's Office of Pipeline Safety so long as they are corrected within five days. Over-pressure events are almost always corrected within this period, but that fact does not reduce the potential harm to the public and the environment that these events can cause by possibly weakening a pipeline. This five-day exemption also precludes PHMSA from getting important safety data that can help identify operators who are having problems properly controlling their pipelines, and that may point to pipeline segments in need of certain inspections. This exemption was removed for natural gas pipelines in the PIPES Act of 2011. Congress should remove the exemption for liquid pipelines as well.

Congress should direct PHMSA to amend its safety-related conditions reporting regulations to require operators of liquid lines to report over-pressurization events.

Prevention Programs

It has been widely recognized among industry and federal regulators that third party excavation is one of the greatest threats to underground pipelines. Pipeline incidents caused by excavation damage can result in fatalities and injuries, as well as significant costs, property damage, environmental damage, and unintentional fire or explosions. While there are regulations which are intended to prevent such damage, such as state "Call Before You Dig" 811 programs, there are still many gaps in these regulations which leave room for the increasing number of excavation related damages caused to pipelines every year.

Under the authority of 49 C.F.R. § 198.35, PHMSA requires that states have a one-call damage prevention system to be eligible for grants from PHMSA to reimburse the costs of its pipeline safety programs. States can receive up to 80% of their costs in grants from PHMSA, but only if they've adopted a one-call system. PHMSA reviews not only the enforcement part of state systems, but the adequacy of the underlying systems as well. Improved enforcement efforts, and PHMSA intervention to provide enforcement when a state won't, may help reduce the number of excavation incidents even further.

While PHMSA has been encouraging states to improve their damage prevention programs, the following concerns continue to come up:

- Exemptions: There are requests for exemptions from participating in the one call system from both the call and response sides of the program. Cities and municipal utilities, state departments of Transportation, and agriculture seek exemptions, or to retain existing exemptions from having to participate in the one call system. Production and gathering pipelines will often seek exemptions from having to participate in responding to one call locate requests or mapping requirements.

- PHMSA maintains that there are no federal exemptions within the Excavation Damage Rule of 2015 and that any exemptions from participating in the one-call system are to be determined at the State level. However, a State must provide to PHMSA a written justification for any exemptions for excavators from State damage prevention requirements. PHMSA will make the written justifications available to the public (§ 198.55(a)(7)).
- Whether an exemption is written as an exception to a definition of what an underground facility is, what excavation is, or whether it's written as an exemption to who must participate, every exemption provides another opportunity for a completely preventable serious pipeline incident to occur.
- Positive response: Not all states require the excavator to be contacted by a utility or the one-call center when all the utilities are done locating and marking. This leads to 2 problems: 1) The excavator is never positive that they've all been marked, even if the 48 hours has passed; and 2) accidents can occur to unmarked utilities even if the excavator did everything right. These issues would be easily resolved by a requirement that the utility either respond directly to the contractor once location is complete, or that the one-call center do so.
- Enforcement authority/equal enforcement: Most state attorneys general have more than enough cases to deal with without adding to their burden by requiring them to enforce violations of state damage prevention laws. Some states have tried to resolve this by creating an independent commission to hear complaints, made up of members from all the various stakeholder groups. This group can hear complaints and make recommendations to an attorney general or a county prosecuting attorney.

Another common concern is that a high percentage of the incidents that cause damage to underground utilities are caused by the utilities being marked incorrectly after one-call has been used. Excavators want to ensure that if they are going to be held accountable for their failures to use the one call system properly, the utilities are also held equally accountable for failures to mark utilities correctly.

Mandate Offshore Pipeline Safety Improvements

Offshore pipeline safety remains an important area for regulatory improvements. These pipelines have unique safety risks and should not be exempt from important safety regulations.

Recent incidents such as the 2020 Enterprise Products propane pipeline explosion in Corpus Christi, TX (4 fatalities, 6 serious injuries) and the 2021 Amplify Energy/Beta Offshore oil spill near Huntington Beach, CA (25,000 gallons of crude oil in San Pedro Bay) demonstrate some of the safety issues with offshore pipelines.

One glaring regulatory shortfall is Offshore pipelines are specifically exempted from having a damage prevention program. Another shortfall is gas pipelines in navigable waterways are not required to have five-year crossing inspections like hazardous liquid pipelines. Congress should address both shortfalls.

Clarify "Confirmed Discovery" Definition

The definition of "confirmed discovery" of an incident is very vague, allowing an operator to potentially delay this notification with little risk of enforcement action by PHMSA.

Pursuant to PHMSA's regulations, operators of gas and hazardous liquid pipelines are required to report incidents to the National Response Center (NRC) and to the Secretary within one hour of "confirmed discovery." 49 C.F.R. §§ 191.3; 191.5; 195.2; 195.52. Unfortunately, this definition of "confirmed discovery" allows an operator to delay notification with little risk of enforcement action by PHMSA. This is delay in reporting an incident can be extremely consequential: Incidents affecting humans or the environment could continue for some time before proper notification and subsequent remedial action begins.

Congress should direct PHMSA to modify its regulations to amend its definition of "confirmed discovery" to ensure that operators notify the NRC and the Secretary as soon as possible after an incident occurs.

Appendix: Statutory and Regulatory Language Where Appropriate

[The appendix to Mr. Caram's written testimony is retained in committee files and is available online at https://docs.house.gov/meetings/PW/PW14/20230308/ 115415/HHRG-118-PW14-Wstate-CaramB-20230308.pdf pages 20–33.]

Mr. NEHLS. Thank you, Mr. Caram.

And I want to thank you all, all the witnesses here, for your testimony. We will now turn to questions from the panel, and I will recognize myself for 5 minutes.

Administrator Brown, the PIPES Act 2020 was signed on December 27th of 2020. Included in the legislation was a requirement that PHMSA modernize its regulations to accommodate the growing LNG export industry. The proposed rule has not yet been published. What is PHMSA's path for the promulgation of these regulations?

Mr. BROWN. Yes, thank you, Mr. Chairman. That rule is a priority rulemaking, one of the many in the 2020 PIPES Act, as far as our monthly updated web chart showing what our priority rulemakings are. And we hope to get a proposal out this year on that.

I was just down in Freeport, Texas, where we had a really terrible incident last year in June. We hope to incorporate lessons learned from that incident in that rulemaking.

I will just note, too, that in the first year of this administration, we had the most productive series of pipeline safety rulemakings in the last 12 years, and maybe beyond that.

So, we are making tremendous progress, but I agree, and I think a theme for the whole day today and the last hearing held regarding PHMSA is that our pace of rulemakings is not what anybody in the room would like it to be. And so, we are excited to work with any and all of you to help improve that ability to get rulemakings done.

Mr. NEHLS. And Mr. Brown, we understand PHMSA has long dealt with staffing challenges, which is why the PIPES Act of 2020 mandated—mandated—that enforcement and inspection personnel not fall below a certain number. Yet as of January 21st, PHMSA has only 207—207—inspection and enforcement personnel, despite the PIPES Act requiring 247 such employees. I think Mr. Grubb was talking a little bit about this, hiring people, safety personnel for 2023.

Instead of filling these essential positions, PHMSA has hired environmental specialists, including an environmental economist, that do not appear directly tied to its pipeline safety mission.

You have also referred to PHMSA as a climate change agency. Considering the agency's current challenges in meeting its pipeline safety mission, do you think rebranding PHMSA as a climate change agency is helpful in recruiting safety and inspection personnel?

Mr. BROWN. Well, so, yes. In fact, when I asked new hires why they came to the agency, a lot of them actually referenced that. Actually CEOs of companies would be interested to hear the representatives here, but the CEOs of oil and gas pipeline companies also cite their focus on climate change over recent years to attract engineers and Ph.D.'s to their agency to be a part of the transition, and to mitigate one of the big challenges.

I will be glad to get you updated numbers on where we are. My understanding is we have actually met PIPES Act mandates in hiring new inspectors and engineers over the last few years.

Mr. NEHLS. OK.

Mr. BROWN. It has been our top focus. We can't do the work without the incredibly dedicated people we have.

Certainly, during the pandemic and over the last few years, the challenges of competing with the private sector, with other States, it is a challenge. I just heard from one of our State agents that they had four people they were interviewing, nobody showed up for the interview. That is real challenging, and any tools you can give us to help us get to that——

Mr. NEHLS [interrupting]. Yes, I appreciate that, Mr. Brown. It has been over 2 years without a nominee for PHMSA Administrator. How has the absence of permanent leadership affected the agency's regulatory efficiency?

Mr. BROWN. Honestly, I don't think one bit, but I don't have too much to add——

Mr. NEHLS [interrupting]. Has there been any discussion when a nominee is coming forward? I mean, when are we going to——

Mr. BROWN [interrupting]. I don't have any estimates or information.

Mr. NEHLS. All right. Last question. And again, in the PIPES Act of 2020, signed on December 27, 2020, included in the legislation was a requirement that PHMSA modernize its regulations to accommodate the growing LNG export industry. The proposed rule has not yet been established.

Again, what is PHMSA's path forward for the promulgation of these regulations?

Mr. BROWN. This door shut when you started. Is this the LNG facilities update rule, the one I think you first asked about?

Our plan is to get it out this year, and we provide monthly updates on progress there.

Mr. NEHLS. Well, I am looking forward to—we had a meeting a few weeks ago. I look forward to continuing on with this relationship so we can do the right thing and make sure that our pipelines are as safe as can be. Thank you.

I will now recognize Ranking Member Payne for 5 minutes. Mr. PAYNE. Thank you, Chairman.

Deputy Administrator Brown, in February of 2020, Denbury's 24- inch carbon dioxide pipe ruptured in Mississippi, with 200 people needing to be evacuated, as was just stated. Fortunately, there were no fatalities, but local first responders were caught completely unaware.

Can you please share with us how PHMSA is using lessons learned from this incident to inform your ongoing carbon dioxide rule development?

Mr. BROWN. Really important question, and I just want to relate it back to the chairman's comments about the LNG facilities rule and discussions we have frequently with industry.

But in that incident that you referenced—and I visited a community of about 70 people in rural Mississippi about an hour north-west of Jackson. The first responders were unaware of the incident and what was at issue with the rupture of that carbon dioxide pipeline. Notification and getting information out—same thing in East Palestine, Ohio.

We live in modern society, where I can get notification of my food order, that it has been retained by the restaurant, they are cooking my food, it is ready to go. That is all within minutes. And in that incident, it was over 11/2 hours before folks were aware that there was a cloud of CO2 gas that was engulfing a community. We can do much better than that.

I really look forward to working with everybody. And I think there is consensus on the need to improve that communication.

Mr. PAYNE. Thank you. As we look forward to a pipeline safety reauthorization, do you feel that PHMSA has the resources needed to investigate all pipeline incidents?

And what additional tools do you need to recruit and retain an inspection workforce?

Mr. BROWN. We triage our needs. I think we have requisite resources to investigate pipeline incidents. But the amount of time it can take, we actually have reduced the amount of time to open, certainly for an enforcement and compliance issue related to an incident, to a record low amount of 323 days.

So, the amount of resources we have—and that seems to be a common theme for the day—is directly proportional to the work that we can do. And everybody seems to think we should be doing more. We are grateful that this committee has continued to support our work. But it is an important question you ask.

Mr. PAYNE. Yes, that is the unfortunate position you are in, irrespective of what you do, people are always going to want more. So, thank you.

Now, Mr. Caram, your testimony says that we are not making real progress on pipeline safety, and that integrity management programs have not lived up to their promise. What are the short-comings of the integrity management plans?

Mr. CARAM. Thank you for the question. Yes, the theory behind integrity management really makes perfect sense. You identify those pipelines that can have the most devastating effects on communities or on the environment in the case of a failure, and you focus your safety efforts on those sections of pipelines with more inspections, more safety requirements, and we identify those areas. We call them high-consequence areas.

But when we look at the data, and we look at incident rates within high-consequence areas and compare them to the incidents outside of high-consequence areas, they are not doing any better. In a lot of cases, they are actually doing worse. And so, we see this as a performance-based metric, a performance-based regulation that leaves a lot up to the operator's discretion. It makes it hard for PHMSA to inspect, hard for them to enforce. And it is not producing the results that we would hope to see, as safety advocates.

Mr. PAYNE. And with the time I have remaining, quickly, what is the single biggest driver of the lack of progress on pipeline safety?

Mr. CARAM. Thank you for the question. It is a good one. I would have to say PHMSA's statutory handcuffs on ability to make strong rules.

Mr. PAYNE. OK. Thank you very much. And Mr. Chair, I yield back.

Mr. NEHLS. Thank you, Mr. Payne.

And Mr. Brown, in your testimony it says PHMSA has kept up with the PIPES Act of 2020 hiring mandates, both for inspectors as well as regulatory personnel. And in your office, January 31st, it says 207 of the 248 inspection enforcement personnel were filled. So, obviously, in the last 5 weeks, 6 weeks you have hired 41 personnel. I find that incredible, with the bureaucracy and the stuff that we have today.

But I will now yield to Mr. Rouzer.

Mr. ROUZER. Thank you, Mr. Chairman.

Mr. Brown, this is not a gotcha question by any means. I am just curious how you will answer it. Generally speaking, are pipelines safe?

Mr. BROWN. Are pipelines safe, generally speaking? Sure.

Mr. ROUZER. Let me ask this. Are investigations common when there is no safety incident?

Mr. BROWN. Can you repeat the question? I didn't hear.

Mr. ROUZER. I am sorry. Are investigations common when there is no safety incident?

Mr. BROWN. Are investigations common when there is no safety incident? Not common, no.

Mr. ROUZER. OK. A followup to that: What circumstances would prompt PHMSA to launch an investigation on a wireline or slickline operator that has never had a recorded safety incident?

Mr. BROWN. Perhaps a notification through anonymous sources or through a hotline. There are many circumstances where you might investigate an allegation of a company or an operator, sure.

Mr. ROUZER. So, let me ask it this way. Is there any kind of testing available for a regulated wireline or slickline operator, where they can be exempt from, quite frankly, unapplicable regulations imposed upon them?

Mr. BROWN. I am sorry, I am having trouble hearing.

Mr. ROUZER. Is there any kind of testing or process available for a regulated wireline or slickline operator to be exempt from unapplicable regulations that might be imposed upon them?

In other words, is there a process where you check all the boxes, you are clear to go, without having to go through an investigation when in fact there is no safety incident?

Mr. BROWN. You would never be totally exempt from—just if you check boxes and you say you are meeting our requirements. We are always going to make sure what you say is accurate, and the operator is held accountable if it is not.

Mr. ROUZER. One other question here. So, what kind of research, if there is any, is PHMSA conducting related to pipelines and the products that support pipelines?

Mr. BROWN. A tremendous amount of research, actually, and it actually gets to Mr. Black's testimony about our commitment to innovation. We spend millions of dollars a year working with the academic community, the regulated community on the needs to improve pipeline safety and technologies. There were actually a bunch of patents come out of the research that Congress appropriates and that we dole out to those communities.

Mr. ROUZER. Is that research available to the public? Mr. BROWN. It is, yes.

Mr. ROUZER. Where can that be found?

Mr. BROWN. Oh, we can send it to you, but it is on our website.

Mr. ROUZER. OK.

Mr. BROWN. Yes.

Mr. ROUZER. So, how many different research projects are there out there?

Mr. BROWN. I want to say at least a dozen. We have got about $8 million currently on some of the new emerging areas of energy— hydrogen, carbon dioxide—both to mitigate safety risks, to improve leak detection, and to keep the product in the pipe, which I think everybody here is focused on. So, that is an area that I think we have a really broad stakeholder support for and engagement with.

Mr. ROUZER. Yes. With regard to my earlier questions, later on I would like for my team to get with your team to talk about some specific incidents, just make sure that you are aware, and that all things are being handled properly.

Mr. BROWN. Happy to. I really appreciate your flagging that. Thanks.

Mr. ROUZER. Thank you, Mr. Brown. I yield back.

Mr. NEHLS. Mr. Rouzer yields. I now recognize Mr. Larsen for 5 minutes.

Mr. LARSEN OF WASHINGTON. Thank you, Mr. Chair. Strengthening pipeline safety protects our communities. We know that in Washington State. It also holds operators accountable. In the future, it can help us fight climate change, with the role of moving CO2 around. I think we need to focus on better identifying our vulnerabilities, addressing safety gaps, and giving PHMSA the tools it needs to strengthen the safety standards.

To that end, Mr. Caram, your testimony outlines a variety of things, but one point you make is that integrity management isn't working. Well, I have been at this 23 years. Many have been at this longer. But it was always about integrity management when we did the Pipeline Safety Act in 2002 and continuing on.

In your view, why isn't integrity management working, and what is the biggest contributor to the pipeline safety challenges that you have identified?

Mr. CARAM. Thank you for the question. I wish I could list off a number of reasons of why it is not working.

Mr. LARSEN OF WASHINGTON. You get to pick one.

Mr. CARAM. I believe it is the performance-based nature of them, and the lack of prescriptions.

For a lot of the—there are a lot of wonderful operators out there that really are leading the way on safety. They take it very seriously. It is a part of their culture. And I would imagine that those operators have pretty good integrity management programs.

Unfortunately, those are not all the operators out there. And for those I think we do need some more prescriptive level of minimum safety standards which would allow the other operators to continue to do their integrity management programs, their safety management systems, and these continuous improvement processes.

Mr. LARSEN OF WASHINGTON. Yes. You have identified a lot of—about—like, 22 recommendations or so in your testimony. I appreciate that.

Are there any new approaches you think would best move the needle to improve safety?

Mr. CARAM. Overall, I think the best thing we can do for safety is give PHMSA the full resources it needs to be the regulator it needs to be. I think they have been historically—it is a theme today among all of us, they have been historically underfunded, they have these statutory handcuffs. And if Congress could give them the tools, the resources, and the ability to become a more robust regulatory agency, I think that is where we will see a real change in pipeline safety, and start to move towards our goals.

Mr. LARSEN OF WASHINGTON. Yes. Mr. Brown, I have been around long enough to not ask you specifically what is going to be in the President's budget tomorrow, because I know what your answer will be. You won't be able to give it.

Should we expect to see something in the President's budget tomorrow with regards to PHMSA and the resources you have outlined that are needed?

Mr. BROWN. Yes, thank you for the question and how you asked it. There are a couple of things that I can highlight without specificity, and one is support for States.

We oversee 3.4 million miles of pipeline, but the vast majority of those are gas distribution, and 80 percent of those are overseen by our State agents. They have gotten about 55 percent of their budgets covered through appropriations that you all provide us to give to them. That is—with all the new demands on us and on them, that is hefty. And so, we are trying to address that in the budget. And then, in addition to that, hydrogen, CO2, the fuels of the future that have greatly been incentivized in the last 2 years through the

Inflation Reduction Act, the Bipartisan Infrastructure Law. The one thing that sort of was missing from those tremendous investments was additional oversight by our agency.

So, the President's budget should reflect those items I can preview for you.

Mr. LARSEN OF WASHINGTON. All right, and so, we will look for those, especially.

Mr. Black, we talked a little bit yesterday about the technology demonstration. Can you talk a little bit more about that?

I mean, it certainly sounds interesting. I want to move forward. Certainly, in the last 25 years, technology has changed. I understand the technology of a smart pig hasn't changed. It is a matter of thinking of that as a platform, and then the stuff you stick on it, if you will, the technology has changed to better do integrity management.

Can you talk a little bit more about the program itself, and how it can benefit?

Mr. BLACK. Sure. Thank you, Ranking Member.

Mr. LARSEN OF WASHINGTON. Yes.

Mr. BLACK. Yes, the smart pig technology or in-line inspection is the same as maybe 20 years ago, but the sophistication of the devices that can travel inside the pipeline looking for issues has improved. So have the analytics of what can be done with that information. We can see more. We have a better idea of what issues might become a problem. So, we know now, through greater advancements, what the proper schedule should be on, which features a pipeline operator should address.

But the PHMSA regulations have not changed that. So, Congress, in recognition, created this section in the PIPES Act 2020, right? That sounded great. But what PHMSA did is they went beyond the mandate and the limitations imposed by Congress.

First, they required that an application to use it follow the NEPA process, and an environmental assessment was done. That is the first time PHMSA has ever done that for an R&D project.

They said the standard of review needed to be higher even than the special permit process, and the special permit is for a permanent waiver of regulations. This is for a temporary project.

And then PHMSA said that anybody asking for a demonstration project needed to use the special permit process itself, which is has taken 10 years, in some instances.

Mr. LARSEN OF WASHINGTON. Thank you. To follow up—and I appreciate the chair's indulgence, and I have nothing to yield back except nothing.

[Laughter.]

Mr. LARSEN OF WASHINGTON. Thank you.

Mr. NEHLS. Mr. Larsen yields, and I will recognize Mr. Stauber for 5 minutes.

Mr. STAUBER. Thank you very much.

Mr. Larsen, thanks for yielding your time to me. I appreciate it. Thank you, Mr. Chair.

In recent years, I am sure everyone has seen the illegal protesters that have set fires, cut hydraulic lines, attacked equipment and operators during the Enbridge Line 3 replacement construction. And they did millions of dollars of damage and put union workers in danger.

Deputy Administrator Brown, are our pipelines, the critical infrastructure used to deliver affordable and reliable energy to my constituents in Minnesota, are they safe from radical protesters?

Mr. BROWN. Congressman, as you and others may know, we don't oversee security of pipelines. So, when you say "safe," our jurisdiction does include the safe operations, but the line is sort of drawn as far as security to the Transportation Security Administration. We do work closely with them. And when we see issues of security, we let them know. And vice versa, if they have issues of security, they flag those issues as they may affect operations.

Mr. STAUBER. When these protesters are protesting for climate reasons, obviously, do they put the workers in jeopardy?

Mr. BROWN. Again, because we focus on the safe operations of the pipelines and not on the security, I am certainly concerned always about the workers doing work——

Mr. STAUBER [interrupting]. But do they put the workers in jeopardy when they jump into pipes as the 49ers are lifting that pipe up into place, does that put them in jeopardy?

Mr. BROWN. I mean, that sounds very concerning to me, sure.

Mr. STAUBER. Does it put the community in jeopardy?

Mr. BROWN. All of that, what you are describing, sounds very concerning——

Mr. STAUBER [interrupting]. Does it put law enforcement in jeopardy?

Mr. BROWN. The way you are describing things, that does sound like it puts everybody in jeopardy if you are threatening violence to anybody or anything.

Mr. STAUBER. Thank you. Mr. Brown, in your testimony you had stated that PHMSA is a climate agency. Do you still stand by that?

Mr. BROWN. Yes. Thanks for giving me the opportunity to explain that.

Mr. STAUBER. Do you sympathize with the radical protesters?

Mr. BROWN. I don't know to what you are referring.

Mr. STAUBER. Protesters, for instance, on Enbridge Line 3 that are climate protesters. Do you sympathize with them?

Mr. BROWN. That is a large group of people who I don't know exactly what you are referring to. Sort of what—I really just don't know what you are referring to. So, I——

Mr. STAUBER [interrupting]. Do you support their radical protests that put the workers, community, and law enforcement in danger?

Mr. BROWN. I do not support anybody putting anybody in danger, especially law enforcement. My dad was a police officer, I come

from a law enforcement family.

Mr. STAUBER. That should have been easier for you to answer, then.

Mr. BROWN. Well, I think I answered the easy ones.

Mr. STAUBER. Are the Biden administration's climate priorities driving decisions at PHMSA, yes or no?

Mr. BROWN. We have a whole-of-Government approach to mitigating the threat of climate change, which I think everybody up here is aligned with.

And I appreciate the bipartisan PIPES Act focusing on minimizing methane emissions, which has 80 times the global warming potential as carbon dioxide.

Mr. STAUBER. Ultimately, I am concerned that this administration is steering PHMSA away from its mission, and using yet another agency to push its activist agenda. PHMSA's mission is to protect people and the environment by advancing the safe transportation of energy and other materials that are essential to our daily lives.

Reliable, affordable, clean energy is good. Would you agree with that?

Mr. BROWN. I would agree that reliable, affordable, clean energy is good.

I think our mission statement also includes protecting the environment, and we also focus on that, as well, and I appreciate the committee supporting that, as well.

Mr. STAUBER. All the American people want is safe, reliable access to affordable energy. We need to protect our pipeline infrastructure to accomplish that goal.

Mr. Chair, I yield back.

Mr. BROWN. Yes, I agree. Thank you, Congressman.

Mr. NEHLS. Mr. Stauber yields. I now recognize Mrs. Foushee for 5 minutes.

Mrs. FOUSHEE. Thank you, Mr. Chairman, and thank you to the witnesses for being here this morning.

I would also like to thank Ranking Member Larsen and Ranking Member Payne for their work, and I am honored to join both of you and my colleagues on this subcommittee.

First I would like to discuss the Colonial Pipeline spill that happened in Huntersville, North Carolina. In August of 2020, gasoline was found bubbling from the ground on the Oehler Nature Preserve in Huntersville. At first, Colonial Pipeline estimated that only 63,000 gallons had leaked. But ultimately, it was reported that nearly 2 million gallons of fuel was released in the nature preserve, resulting in my State's largest gasoline spill ever.

Last summer, the North Carolina Department of Environmental Quality filed a consent order holding Colonial Pipeline accountable after they failed to provide the State with essential information required for remediation after the spill. It took 2 years and a consent order.

Mr. Brown, in May 2022, your agency proposed nearly $1 million in civil penalties on Colonial Pipeline for multiple probable violations of Federal pipeline safety regulations for failing to adequately plan and prepare for a manual restart and shutdown operation, which contributed to the national impacts of the May 2021 cyber attack.

So, my question is whether PHMSA's civil penalty authority is substantial enough to encourage better behavior by pipeline operators.

Mr. BROWN. That is a really good question. I don't know if I have a great answer. I would welcome other panelists' answer of whether a civil penalty of $1 million is going to dissuade a company that is valued in the billions. Intuitively, I would think not.

But I will say, we specifically highlighted that proposed civil penalty because it was something where we had actually given the company a heads up—hey, you have got to fix this—and they chose not to after we even told them to adhere to our regulation. For us, that is a particularly egregious violation, and that is why we issued a proposed civil penalty.

Mrs. FOUSHEE. Thank you for that response.

Mr. Grubb and Mr. Black, I am interested in learning more about how you detect and prevent leaks in your industry. Is this a priority for Kinder Morgan, Mr. Grubb?

And for your member organizations, Mr. Black, do you have a means of detecting leakage on your pipelines, and are your systems for doing so more stringent in high-population areas, also known as high-consequence areas?

Mr. Grubb, you may go first.

Mr. GRUBB. Thank you for the question. So, in terms of detecting methane leaks, we routinely patrol our pipelines for leak detection purposes. We also implement leak detection at compressor stations along our pipeline routes.

Detecting leaks is important. It is something that does not happen often, in terms of natural gas pipeline leaks. But when it does happen, we quickly identify it, and we quickly repair those.

Methane emissions, in general, are something that is of the utmost importance to us, as well as our industry. And certainly PHMSA acting upon the rulemaking to address class location will also minimize or reduce methane emissions associated with activities that are required to meet the current regulations, that being pipe replacement. And it will minimize the unnecessary release of natural gas associated with those projects, and that will go a long way in reducing methane emissions, as well.

Mr. BLACK. Thank you, Congresswoman. Detecting leaks is very important. A pipeline operator's goal is to detect the leak available in the control room, where you can shut down the pipe at the sign. So, they maintain sensors along the pipeline, looking for signs of a leak.

Industry, in the last several years, has developed through experts a recommended practice at the American Petroleum Institute on leak detection. We are also conducting R&D to continue to improve leak detection. We want to be able to see, smell, or hear maybe the signs of released product, also maintain those aerial patrols, just like Mr. Grubb was saying, and have community awareness programs so that, if the public knows—if they find an incident of a leak, they know who to call. Detecting a leak is very important.

Mrs. FOUSHEE. Thank you.

With that, Mr. Chair, I yield back.

Mr. NEHLS. Mrs. Foushee yields. I will now recognize Mr. Johnson for 5 minutes.

Mr. JOHNSON OF SOUTH DAKOTA. Thank you, Mr. Chairman. I was grateful to see that Ranking Member Larsen had focused a little bit on the technology pilot program.

And so, Mr. Brown and Mr. Black, I want to dive a little deeper. I mean, talk to me about what improvements we could make to the program so that it would be more valuable for operators, and more people would take advantage of it.

Mr. BROWN. You want to go? You want me to—go ahead.

Mr. BLACK. OK, thank you. Nobody has used the program right now. It has been untenable. We need to make some changes.

One, we need to change the standard of review. It is complicated in the law. It is complicated in what PHMSA has done. Generally, we want something simple, where it is a reasonable level of safety, compared to what the existing regulations have.

Second, we don't——

Mr. JOHNSON OF SOUTH DAKOTA [interrupting]. So—hold on. Before going to second, let's drill in just a little bit. So, when you say we need to change the standard of review, what is it, and where do we need to head to?

Mr. BLACK. The problem is that PHMSA is using a standard of review even higher than its special permit process, which is for permanent waivers of regulations. It doesn't need to be that way.

Shall I move to another one?

Mr. JOHNSON OF SOUTH DAKOTA. What standards should they move to? Should they just move to their standard level of review?

Mr. BLACK. We want PHMSA to exercise judgment on if this application is going to improve pipeline safety.

Mr. JOHNSON OF SOUTH DAKOTA. I see.

Mr. BLACK. This is PHMSA's decision to make. They have made it more complicated than they need to.

Mr. JOHNSON OF SOUTH DAKOTA. OK, very good, sorry. Number one. Now number two.

Mr. BLACK. Great. Next, we need to not use the NEPA process. There is no reason for an environmental assessment to be completed for an R&D project.

Next, we need to not use the special permit process itself. It is broken. Everybody recognizes that. It has taken 10 years for some special permits.

Next, we need to extend this program. Congress gave it a few years. We have lost 2 years already. Do a 10-year reauthorization so we have the full cycle of PHMSA to finish developing the updates, operators to apply. Let's conduct the research. PHMSA can get the data. The goal is for them to update their regulations with the know-how of 20 years of developments.

Mr. JOHNSON OF SOUTH DAKOTA. That is excellent, because those are specific changes that we can make that are within our power, they are easy to understand, and they would have the desired impact.

Sir, what did he get wrong?

Mr. BROWN. So, we actually opened for public comment, "How can we implement this directive from Congress?" We didn't get the specifics of exactly your question, Congressman. What standard do we judge a new technology—it is just an R&D project. That is different than deploying a technology that has safety implications for the public. I rely on the subject matter experts, whose focus is safety, to make that determination.

Now, the question of how to apply the National Environmental Policy Act, that is a directive that Congress established 50-plus years ago. You can consider environmental impacts in different ways, but there was no proposal to provide those environmental impacts at all in the public comment period that we offered.

We want innovative technologies. We are happy to work with folks on this. That is why we opened it up for public comment. Happy to work with LEPA and any of the committee members here. But we don't want to sacrifice two things, right? Safety, environmental impacts. So, how to do it is absolutely the question. You asked the right question. We are asking the same question. And if we can come to agreement on an efficient way to do it, great. It is definitely a common goal that we have.

Mr. JOHNSON OF SOUTH DAKOTA. Mr. Black, what else?

Mr. BLACK. PHMSA has recognized that a decision on a pipeline safety demonstration program application is not a major Federal action. It has the discretion to not put it through the NEPA process. This can be simple. Congress can tell PHMSA, "Don't go beyond the limitations that we tell you to use in this program."

Mr. JOHNSON OF SOUTH DAKOTA. Certainly——

Mr. BROWN [interrupting]. Well——

Mr. JOHNSON OF SOUTH DAKOTA [interrupting]. No, go ahead, sir.

Mr. BROWN. So, we certainly, potentially, could face a lawsuit, and then have to litigate that. If Congress were to provide us the directives, or if we could get a proposal that says we are not going to ignore the environmental impacts, here is how you know we are not going to ignore the environmental impacts, then that would get us closer to what I think we all want to get to.

Mr. JOHNSON OF SOUTH DAKOTA. So, Mr. Chairman, I would just close by saying this. I mean, we have specific recommendations. We have an opportunity to make this program work better, to get additional research and development, to make the system run better and more safely. It is not being used properly. This is something we can work on together. This is something that we might be able to get the overwhelming majority of the people on both sides of this dais to understand that the reforms that have been described will work, and they will make the system better long term. Let's get it done together. I yield back.

Mr. NEHLS. Mr. Johnson yields. I now recognize Mr. Cohen for 5 minutes.

Mr. COHEN. Thank you, Mr. Chair, and I thank you and the ranking member for holding this hearing.

The safe operation of the 3.4 million miles of pipeline regulated by PHMSA is vital to the health and safety of the American people. I want to focus on a pipeline issue that is particularly important to our country's ability to combat the climate crisis, which is methane leaks.

According to a recent report in a Guardian article, methane causes about one-quarter of global heating from the greenhouse gas effect, and human-caused emissions over the last century have contributed about one-third of the total global temperature rise. Emissions have surged in recent decades, and if we continue on this path, methane alone could push us past a number of catastrophic climate tipping points.

Around 40 percent of those methane emissions come from leaks from fossil fuel exploration, production, and transportation, and that percentage has increased by half in the last two decades. For the sake of the planet and future generations, we must do all we can to prevent methane leaks from the millions of miles of natural gas pipelines under PHMSA's authority.

In my State, the Tennessee Valley Authority is pushing forward with a plan to construct new natural gas powerplants and pipelines as a "bridge" to renewables, a bridge to renewables. I have argued that it is a bridge to more pollution and disasters, and have strongly encouraged them to rethink this approach.

Deputy Administrator Brown, in your written testimony, you stated that PHMSA has been integral to the whole-of-Government approach to mitigating unnecessary greenhouse gas emissions. Could you expand on what this approach entails, and what resources or support PHMSA might still need to properly carry out this vital mission?

Mr. BROWN. Yes. Well, thank you so much, Congressman. And really, we have been really energized by this committee's work with the 2020 PIPES Act, which said regulated facilities have to minimize methane emissions—as noted, 80 times the global warming potential as carbon dioxide.

Also, we have to include in our cost-benefit analysis, as we move forward with rulemakings, environmental impacts, which just gets to some of the questions from Members today why we might hire people with environmental knowledge and skills. We have been directed to do that by the committee in carrying out these directives. And we are doing it, actually, by hiring engineers with environmental knowledge and experience.

But we have got six rulemakings in the White House Methane Emissions Reduction Action Plan, three of them we finalized. And those overlap to the chairman's questions about how we are prioritizing. We are prioritizing the directives we have from Congress to do a host of rulemakings, the ones that have the biggest safety impact, that—also, I think it is inextricably linked. You reduce methane emissions, you are actually improving and mitigating climate change. You are mitigating methane emissions when you keep the product in the pipe, which is something you hear from industry all the time.

We are all, I think, focused on that, and how we talk about it. And I suggest we talk about it. I suggest we talk about keeping the product in the pipe is helping reduce methane emissions and global warming. The agency that I lead now wasn't allowed to talk about that for 4 years. And a lot of companies are talking about it now to recruit folks, and also because their investors want that information.

So, really, over the last few years, the marketplace, the free market has said, yes, actually we would like to have improved clean energy, but we also want to mitigate the impacts to the environment.

Mr. COHEN. Our time is running out. If I can—I thank you for that—very quickly, could operators decide for themselves which high-consequence area definitions to use?

Why do we have two different definitions, and do we allow operators to pick and choose even different ones along the same route?

Mr. BROWN. I am sorry. I didn't hear that.

Mr. COHEN. Well, I got 1 minute left. I am going to pass.

Mr. Caram, in your written testimony, you state that natural gas regulations drafted before the collective consensus that methane emissions are a major contributor to climate change are extremely permissible with regard to leak reporting requirements.

You and Deputy Administrator Brown, methane, a potent greenhouse gas, we want to avoid it leaking from pipelines. PHMSA has just finished a year-long study to put together a proposed rule on the use of advanced leak detection. Where are we tackling methane emissions from natural gas pipelines, and what still needs to be done?

Mr. CARAM. Thank you for that question. As Mr. Brown stated, Congress made extraordinary progress in this area with the PIPES Act of 2020, but we still do have a lot of work to do. We now have satellites detecting huge methane plumes from planned and routine maintenance from pipelines. We don't tolerate these kind of planned intentional releases from liquid pipelines, and we need to continue this culture change on the gas side, and no longer tolerate those, as well.

These blowdowns, the intentional releases for routine maintenance, they are not required to be reported to PHMSA. And besides the general language in the PIPES Act about minimizing emissions, there is no limitation on them.

Mr. COHEN. Mr. Brown, do you have something to add?

Mr. BROWN. Yes, we have a rulemaking that is at the Office of Management and Budget that addresses leak detection, repair, blowdowns, basically what the gentleman just addressed, Congressman, and that is a mandate from the 2020 PIPES Act.

Mr. COHEN. Thank you, sir, and I yield back the balance of my time.

Mr. NEHLS. Thank you, Mr. Cohen. I now recognize Mr. Yakym for 5 minutes.

Mr. YAKYM. Thank you, Mr. Chairman. And thank you to our witnesses for taking the time to be here today and share with us your expertise.

I am glad that our Transportation and Infrastructure Committee kicked off our first hearing on the FAA reauthorization with a hearing concerned and centered around safety. And I am glad that we are doing the same thing with PHMSA today, with the reauthorization, a hearing on safety.

Pipelines are the lifeblood of our economy. They help put gas in our vehicles, they heat our homes, they power our factories and our businesses, and they drive exports that lift our economy and create jobs right here at home. They are an integral part of the conversations as we discuss emerging technologies like hydrogen and carbon capture, as well.

At the heart of all these applications is safety. Without pipeline safety, Americans won't have gas, heat, power, or exports that we depend on to drive our economy on a daily basis. That is why we have agencies like PHMSA, the Pipeline and Hazardous Materials Safety Administration, which is entrusted with this critical mission of safety.

But I am concerned—concerned because it seems like there is a potential movement to deviate from that mission of safety. We heard earlier today about a number of pipeline issues that have happened, safety issues across our country, and it seems like we may have a deviation from that mission of safety.

Mr. Brown, you recently said, and I quote, "One thing we are trying to brand is as a climate agency." I would like you to clarify that comment, because Congress created PHMSA to be a safety agency, not a climate agency.

Full stop. So, if you could for the record, Mr. Brown, please clarify. Is PHMSA's top and sole priority safety?

Mr. BROWN. Yes, it is. It is the top priority, but it is inextricably linked to environmental impacts. And that is in our mission statement, and it is in the 2020 PIPES Act for us to minimize methane emissions, methane being 80 times the impact of carbon dioxide in global warming. When you minimize methane emissions, you are mitigating climate change.

Now, you can talk about it, or you can not talk about it. I choose to talk about it. But when you minimize methane emissions—and by the way, most of the companies that Mr. Black represents, Mr. Grubb, his company, they talk about what they are doing to mitigate climate change. It is inextricably linked to safety. And so, to just not talk about it, I mean, it is up to folks if we want to talk about it. I choose to talk about it. I think it is good for recruiting. It happens to be accurate. And we are acknowledging the reality that climate change exists, and we have to consider ways to mitigate it.

Mr. YAKYM. Do you think that some of the pipeline safety issues we have had have happened because maybe we have taken our eye off the ball of safety, and we are focused more on climate change, it seems, than we are on——

Mr. BROWN [interrupting]. Absolutely not.

Mr. YAKYM [continuing]. Safety, which is the responsibility of your agency?

Mr. BROWN. There are 600 people completely dedicated to safety. You couldn't convince them not to be focused on safety. The first thing I did was what are the rulemakings that will have the biggest safety impact at the agency, and that is what we are focused on.

Mr. YAKYM. Talk to me about how your agency speaks to recruits. I mean, do you tell your recruits that you are a climate agency that is concerned about safety?

Mr. BROWN. I mean, as far as what we recruit, we post our jobs. I don't think most of them mention climate change. But through our social media, through hearings like this, we are trying to note, hey, if you want to make a difference—and when you talk to people under the age of 30, 35, they really care about making a difference in the world and leaving something better for the next generation. It is different than what people in the 35 to 45 care about, or the 55 to 65.

We want to recruit from the biggest pool of people that we possibly can to get the best engineers, best safety experts. And if that can provide a little bit of difference, and get us a little bit more folks to address the chairman's comments from earlier of how to recruit people, great.

Mr. YAKYM. How would you rate the considerations in hiring when you are considering between safety considerations in hiring versus climate considerations in hiring?

Mr. BROWN. Well, for the vast majority of our positions, safety is integral. We have a few positions—we started off with less than 1 position—at an agency of 600 people who has in its mission statement environmental impacts, we had less than 1 focused on National Environmental Policy Act implementation with respect to our agency. We now have three and change, one person who is straight out of college, right?

I mean, we have not revamped our agency of 600 people with 3 people. We just now are able to carry out the laws that you have given us to carry out, which include environmental impacts.

Mr. YAKYM. All right, thank you. And Mr. Chairman, I yield back.

Mr. JOHNSON OF SOUTH DAKOTA [presiding]. Thank you.

With that, Mr. Garc´ıa, your 5 minutes.

Mr. GARC´IA OF ILLINOIS. Thank you, Mr. Chairman and Ranking Member, for holding this hearing. This committee, of course, has a responsibility to ensure the safety of our transit and infrastructure systems, and this hearing is critical to ensuring that we provide that safety.

There is currently a proposal to build a 1,300-mile pipeline to transport liquefied CO2 across four Midwestern States, with Illinois serving as the endpoint and storage facility. The building of this pipeline will cause significant safety concerns. And to date, there are no plans from PHMSA to update CO2 safety regs until 2024. Mr. Brown, regulators are behind on all these issues. And given past mismanagement, why would we trust corporations to come up with their own appropriate safety standards?

Mr. BROWN. I don't think we should trust anyone. I think we should put rules in place, and standards in place, regulations in place that are foolproof, that don't involve trusting, but involve meeting standards.

And we are planning to promulgate a draft rule in the coming months. I think that will set the standard and bar for the projects you referenced.

Mr. GARC´IA OF ILLINOIS. And as the pipeline goes, shouldn't we hold off on construction?

Mr. BROWN. That is not a question that Congress has given to my agency to make any sort of determination.

I can just tell you we are racing to get these rules out, to make them as strong as possible. We have got input from, I think, folks across the panel here to acknowledge and the consensus that there need to be updates to the rules.

Mr. GARC´IA OF ILLINOIS. Fair enough. And my next question, Mr. Brown, is this: If we don't have definitive standards for how far plumes go, or how far back setbacks should be, how can we accurately regulate for the health and safety of those nearby?

Mr. BROWN. That is something we hope to address in our rule-making.

Mr. GARC´IA OF ILLINOIS. Why would we put this technology in place, risking lives without adequate research or protections in place?

Mr. BROWN. Well, so, just for clarification, we have had carbon dioxide pipelines around for decades. And it was incentivized in the bipartisan infrastructure act as well as the Inflation Reduction Act, because of the climate change challenge that we face, and this being one of the potential solutions to the challenge that I think— I hope everybody in the room can help us focus on, which is how to make sure it works, how to make sure it works safely because of its potential to reduce climate change impacts.

Mr. GARC´IA OF ILLINOIS. Thank you.

Mr. Chair, I yield the balance of my time.

Mr. JOHNSON OF SOUTH DAKOTA. Thank you very much, Mr. Garc´ıa. With that, the gentleman from Texas, Mr. Babin. You have 5 minutes.

Dr. BABIN. Thank you, Mr. Chairman. I appreciate it. Thank you, all of you witnesses, for being here.

Pipelines are the backbone of our energy infrastructure. The United States has the largest network of energy pipelines in the world. My home State of Texas has the highest total pipeline mileage by far. And while there is certainly room for improvement, studies prove that pipelines are also the cheapest, the most environmentally friendly, and the safest way to transport oil and gas and other products.

However, given the Biden administration's actions to shut down pipeline projects like the Keystone XL pipeline that they killed back in 2021, you would think that it was just the opposite. Why would President Biden shoot down a cheaper, cleaner, safer way to move our petroleum products, a project that supports our national security and energy goals? It is a great question, and one we ought to be asking ourselves every single day.

But moving forward, we have got to be sure that the Pipeline and Hazardous Materials Safety Administration is fully equipped and prepared to oversee current and future pipelines. And they also need to be laser-focused on safety, not environmental advocacy or anti-corporate litigation efforts.

So, Deputy Administrator Brown, due to the growing size of the liquefied natural gas, or LNG, market and its importance for both domestic and international relations, PHMSA will need to step up its important safety role to oversee this industry.

One proposal is that PHMSA institute an LNG separate Center of Excellence at a new location, as discussed in its report to Congress under model 3. Model 3 appears to provide comprehensive benefits to directly align industry, stakeholders, and industry in the continued growth of the domestic LNG export facilities. If PHMSA does not intend to pursue model 3, how will it make up for the shortfalls and gaps in knowledge that would occur?

Mr. BROWN. This is with regards to the creation of the LNG Center of Excellence?

Dr. BABIN. That is model 3, that is what—I think you mentioned that in your opening statement, did you not?

Mr. BROWN. Yes, I don't think model 3 was in my opening statement, but——

Dr. BABIN [interrupting]. OK, well——

Mr. BROWN [continuing]. But we did get appropriated $8.4 million just a few months ago to create the LNG Center of Excellence, and so, we are well on our way in developing that model, but don't have anything to report yet as far as where, and some of the specifications of what that will look like. But it is something that—we are grateful that Congress has appropriated this funding to help improve the safety and environmental performance of LNG facilities.

Dr. BABIN. OK, well, I have got another question, time permitting. The Federal permitting process for energy infrastructure projects is complex, often triggering the jurisdiction of multiple Federal agencies and reviews. This is especially true for large-scale LNG projects, where PHMSA provides the regulatory expertise on safe design and operations. Yet FERC holds the authority to authorize the construction of these projects.

I understand that often, because of a lack of coordination between the agencies, project developers, despite going through a multiyear design process, are then questioned and then forced to redesign existing facilities to satisfy the evolving demands of Federal inspectors and regulators.

What is your agency doing to improve this coordination and to help shorten the permitting process, so that some of these projects can move forward, especially in an era of shortfalls of energy?

Mr. BROWN. Yes, and thank you, an important question. While we don't have, as you recognized, permitting authority, we do work closely with the Federal Energy Regulatory Commission. I was actually just down in Freeport,

Texas, the site of an incident last year, a 450-foot-high fireball and explosion at that facility. We want to learn from those mistakes, and make sure we don't have that type of incident ever again.

And so, we had, for the last 9 months, met on a weekly basis with the Federal Energy Regulatory Commission, as well as the U.S. Coast Guard, to basically not, sort of, stay in our lanes, but instead work more closely, as you describe, to identify any potential areas of risk, whether it is in our lane or their lane typically, and work more robustly to identify those items upfront in that process.

Dr. BABIN. OK, Mr. Chairman, I yield back. Thank you.

Mr. JOHNSON OF SOUTH DAKOTA. Thank you, Mr. Babin. We now turn to the new Member from New Jersey, Mr. Menendez.

Sir, you are recognized for 5 minutes.

Mr. MENENDEZ. Thank you, Mr. Chairman, and I thank my friend, Ranking Member Payne.

Deputy Administrator Brown, as you know, we have seen increasingly sophisticated cyber attacks on our transportation infrastructure, including our pipeline systems. Keeping our pipelines secure is critical to the safety and well-being of the communities we represent. Can you speak to some of the cyber threats PHMSA grapples with, and what the interagency process for dealing with these attacks looks like?

Mr. BROWN. I can speak a little bit, but just for everybody's benefit, Congress has not given us authority to oversee cybersecurity of pipelines. The Transportation Security Administration does do that. But we do work very closely with them.

Prior to the Colonial Pipeline cyber attack, we would engage voluntarily, invite our regulated entities, pipeline operators to discuss their cyber plans because, as we experienced when that incident occurred, there are operational impacts, potentially, when you have a cyber attack, and we are responsible for safe operations. Ultimately, the companies are always responsible for safe operations, but we are responsible for overseeing that. And so, prior to that incident, we had about 50 percent of our regulated entities engage in those discussions. Post that incident, we had about 100 percent of entities engaged in that.

So, we have provided our input to the Transportation Security Administration on their proposed security directives on cybersecurity. We have engaged with leadership of pipeline companies, including the CEOs of companies that Mr. Black and Mr. Grubb represent, to make sure we are all on the same page. And more coordination between the Government and the private sector and between the Government agencies like the ones I referenced is integral to mitigating this increasing threat.

Mr. MENENDEZ. And a quick followup to that. Do you believe you have the statutory authority and tools you need to keep our pipeline infrastructure secure?

Mr. BROWN. We don't have the security responsibility, but we are actually hiring a cyber expert, even though we don't have a jurisdiction in that space, it does affect the areas that we do have jurisdiction.

The nexus between operations and cybersecurity, just the same as Congressman Stauber noted for physical security. When we find issues, we want to flag them for Transportation Security Administration. And when they find them, we want to know about that. We train their inspectors on pipeline safety issues. So, we will continue to keep up that close collaboration.

Mr. MENENDEZ. That is great, and I appreciate that. And speaking of collaboration, this next question is open for all the witnesses. Where do you see the gaps in our pipeline cybersecurity infrastructure?

And what do you need from Congress to address these threats?

Mr. BROWN. I will answer last because I have talked a lot here.

Mr. BLACK. TSA has primary jurisdiction, as the Deputy Administrator has said, and they have identified critical facilities and required security directives of pipeline operators. That engagement with TSA and DOE is pretty good. Pipeline operators participate in a public-private partnership, Oil and Natural Gas Sector Coordinating Council, where we can do classified briefings and talk threats.

TSA has seen that they need to evolve their requirements on pipeline operators to make sure that we are keeping up with the latest threats. There are a lot of bad guys out there who are trying to stop pipelines, but the public-private partnerships and the actions by pipeline operators are heading in the right direction. I don't have a recommendation for you.

Mr. MENENDEZ. For sure. Especially, you see more sophisticated technology, in terms of the pipeline management. I can imagine the increase in technology also increases the exposure and the risk. And so, I appreciate you continuing to monitor with our different agencies, TSA being the primary lead.

Mr. BLACK. Yes.

Mr. GRUBB. So I, as well, don't have a recommendation, but I appreciate the question.

But I will tell you that Kinder Morgan is definitely engaged with TSA and CEPA regarding cybersecurity of our pipelines. And in fact, our CEO was in

Washington late last week, meeting with those agencies and talking through, and collaborating on the importance of cybersecurity and plans and programs going forward.

Mr. BROWN. I will just add one thing we have floated is having a detailed firm, a regulated entity with cyber expertise. Again, while we don't have that jurisdiction, we would still benefit, I think, from the sort of internal knowledge of how an operator navigates some of these challenges and vice versa that I think they can appreciate what we need to know in an incident like the Colonial cyber incident.

Mr. MENENDEZ. Thank you. I appreciate all of you, and I apologize for not having any time to yield back.

Mr. NEHLS [presiding]. Thank you. The gentleman's time is expired. I now yield 5 minutes to Mr. Duarte.

Mr. DUARTE. Hello. Thank you, Mr. Chair, Ranking Member. I appreciate the hearing today.

Mr. Brown, good day. The fiscal year 2023 Appropriations Act provided more than $319 million in new funding for the Pipeline and Hazardous Materials Safety Administration. And as you know, the PIPES Act required PHMSA to maintain a minimum of 247 pipeline inspection personnel for fiscal year 2023. Your current count is 207, and now we learn that this funding includes a total of 33 positions, including 6 positions for pipeline safety climate change experts.

Given the clear direction of Congress that PHMSA hire a minimum of 247 pipeline inspection and enforcement personnel, how can you justify 6 positions related to climate change in an agency that is supposed to be focused on safety?

And what are these six climate change experts doing that is not covered by the regular pipeline inspection personnel?

Mr. BROWN. Yes, I want to make sure we get a very robust answer as a followup to this hearing, because over the first 2 years, we were meeting the year-on-year increases of 20 percent each year. At the end of the year, we often lose folks to retirement. I want to make sure we get you up-to-date numbers and factor in those changes.

We also initiated new recruitment and retention proposals that are currently pending with the Office of Personnel Management. For example, the Bureau of Safety and Environmental Enforcement pays their engineering inspectors about 30 percent more than we are able to pay ours.

Additionally, there is at least one State that pays about 30 percent more. We just want to be able to be on that same level playing field.

We can't compete with the private sector in terms of compensation, but we welcome any incentives.

Mr. DUARTE. Great.

Mr. BROWN. It is not for a lack of trying. We certainly want——

Mr. DUARTE [interrupting]. Well, thank you. My followup question is, what are the climate change experts

Mr. BROWN [interposing]. Yes, great.

Mr. DUARTE [continuing]. Doing that your regular safety personnel are not doing?

Mr. BROWN. Yes. So, this committee directed us to consider environmental impacts in our rulemakings. That has been a directive from this committee. We have to implement that.

The last administration promulgated a rule in which 15 State attorneys general sued us because we did not consider environmental impacts. When a rule is litigated and basically held in abeyance, we can't do our work on those rules until we consider the environmental impacts.

As I noted before, we had less than one full-time employee considering environmental impacts when it is in our mission statement. So, we need a few folks. That also includes lithium-ion batteries. So, half of our agency oversees 1 in 10 goods that move in the United States: nuclear waste, lithium-ion batteries——

Mr. DUARTE [interrupting]. OK, I got it. It has been said that if everything is a priority, then nothing gets priority. PHMSA requests more resources from Congress to complete this core mission. Now you are spending these resources on activities that have no correlation with pipeline safety.

We can all talk about climate change until the cows come home. We can all talk about: Are we facilitating the use of natural gas, and does natural gas contribute to climate change? Your core mission is keeping the natural gas and the natural gas pipeline—because as you said, it is a very powerful—when released, a very powerful climate driver.

And so, in this change in mission—is this change in mission being directed by the White House? And what communications have you had with them regarding this new direction for PHMSA?

Mr. BROWN. So, I agree with what you said. So, I think maybe there is just a disconnect on, sort of, our common understanding. When we are keeping product in the pipe, that is mitigating climate change, right? So, that is what we are trying to do, is keep the product in the pipe. That is improving safety, mitigating——

Mr. DUARTE [interrupting]. I guess I have to ask. You used to have quarterly meetings with the industry, the PHMSA group and your group used to have quarterly meetings. And yet, the best way to keep gas in the pipe and to keep environmental catastrophes from happening is working closely with the people who are most engaged in keeping them from happening, as the industry folks here are very incentivized to deliver safety and delivery within their pipelines. You haven't been having the meetings since 2021. How can you achieve your stated purpose without better engagement with the industry?

Mr. BROWN. I believe we have had five advisory committee meetings in the last few years. Those are primarily focused on—and their role as Congress set out—to consider rulemakings. We had the most productive year in our agency's history in 2021 in rulemakings. Many of those had already had the advisory committee meeting for those rulemakings.

So, you can expect a bunch more, either a bunch more meetings or longer meetings to just get more done at one meeting, instead of having a bunch of separate meetings, because it takes a ton of staff time to actually set up for those meetings. And it also takes a lot of time—and I know people are very generous, we have got members here represented in actually going to those meetings, but we are trying to make it as efficient as possible, instead of just having meetings, which in this town often don't always produce results.

Mr. DUARTE. I can see that. I will yield back.

Mr. NEHLS. Thank you. I now yield 5 minutes to Mr. Carter.

Mr. CARTER OF LOUISIANA. Thank you, Mr. Chairman and Ranking Member—273 crossing fatalities, 1,475 per year between 2005 to 2021. GPAC meetings, making sure these communities are safe, becomes very, very important.

Mr. Grubb, can you explain the benefits of the Pipeline and Hazardous Materials Safety Administration issuing a final class rule?

Mr. GRUBB. So, it has three main benefits. I thank you for the question, by the way. There are three main benefits associated with finalizing the class location rule.

The first is updating the outdated regulation that has been in place for over 50 years. It was put in place prior to in-line inspection being developed, and those technologies being further advanced.

The second benefit is, we will eliminate over 800 million cubic feet that is emitted annually by virtue of having alternative methods of compliance with the current regulation.

The third benefit is the hundreds of millions of dollars that industry is spending on replacing perfectly safe pipe that is in accordance with the current regulation, can be reallocated to other safety measures, other safety benefits on these pipeline systems.

We hope that PHMSA will finalize this rule as soon as possible, based on those three benefits.

Mr. CARTER OF LOUISIANA. Can you share with me what your company does to ensure pipeline safety?

Mr. GRUBB. So, the main things are, we respond to incidents with the priorities of lives and property, first. We investigate incidents to determine the causes and the facts of those incidents. And then we use those findings to improve our processes and procedures.

Mr. CARTER OF LOUISIANA. So, we know how important it is to involve the community. And I just heard Mr. Brown suggest that there have been five meetings over the last several years. Do you think that is ample?

I mean, we are talking about keeping the community involved, making sure that they understand the most up-to-date safety measures, that they know the early warnings, that they know their area is to be protected. I suggest that these meetings should be much more frequently held.

Mr. BROWN. These are largely to consider the text of regulations that we promulgate, and come to a consensus on that text. And so, we are trying to time them where we can get—given the interest in our finishing our rulemakings quickly—as many rulemakings on one of those meetings as possible. Otherwise, it extends the amount of work everyone has to do.

Mr. CARTER OF LOUISIANA. But you would agree that having community involvement is essential?

Mr. BROWN. I am sorry?

Mr. CARTER OF LOUISIANA. You would agree that having community involvement is essential?

Mr. BROWN. Yes. Community is welcome to attend those meetings, but it is largely only attended by the statutory members of that committee, which is 15 individuals, and they represent 5 people from industry, 5 from governmental entities, and then 5 public members.

Mr. CARTER OF LOUISIANA. What individuals are mandated that must be in attendance for these meetings?

Mr. BROWN. Just the—well, actually, no one is mandated to attend, but there are 15 members of the committee that are statutorily required to be appointed.

Mr. CARTER OF LOUISIANA. Fifteen members from where?

Mr. BROWN. Five from industry, five from the public, five from governmental entities, State and local.

Mr. CARTER OF LOUISIANA. OK, so, tell me about the five from the public.

Mr. BROWN. So, we, for the first time, appointed a labor representative, the people that actually do the work that we are talking about on the ground. Mr. Caram is a member, and his organization is represented as a member of the public. We have an environmental organization, Environmental Defense Fund, represented, a law professor, and then, for the first time, we are trying to focus on, can we get some engineer professors, or people with sort of the greatest amount of expertise in this space committed to our mission of safety and environmental——

Mr. CARTER OF LOUISIANA [interrupting]. Well, why not individuals? Why not people from the community, the people that are most impacted? You say those five are community, but I didn't hear you say anybody from a neighborhood. I didn't hear anybody from a school, anyone from a church, anyone from the affected community. May I——

Mr. BROWN [interrupting]. Yes, we do.

Mr. CARTER OF LOUISIANA [continuing]. I——

Mr. BROWN [interposing]. Yes.

Mr. CARTER OF LOUISIANA. Let me ask first.

Mr. BROWN. I am sorry.

Mr. CARTER OF LOUISIANA. It is important that, when we talk about these things, that we involve community.

Mr. BROWN. Absolutely.

Mr. CARTER OF LOUISIANA. And if you call it a community five, then that community should truly be representative of the people in the community.

Mr. BROWN. My understanding of what Congress——

Mr. NEHLS [interrupting]. The gentleman's time is expired. I now yield 5 minutes to Mr. Molinaro.

Mr. MOLINARO. Thank you, Mr. Chairman, and I appreciate that. Mr. Brown, I want to get to you in a moment, but I strongly suggest that long meetings don't produce better outcomes. Effective meetings produce better outcomes. Congress is a living example of some of that, and I trust that you can appreciate that.

I do want to, Mr. Grubb, though, I want to follow up on the classification rule. I appreciated my colleagues' questions. Could you remind us, however, how long the industry has been waiting for the rule to be finalized?

Mr. GRUBB. Thank you for the question. Over 20 years.

Mr. MOLINARO. So, Mr. Brown, what exactly is the reason why over 20 years is the time in which the industry has been waiting?

Mr. BROWN. The question the chairman asked me first about the LNG facilities rule, that is about 30 years. That one, big safety impacts. This one, more efficiency, potential impacts for reducing environmental emissions. We focus on safety first.

We had three rules that we finished last year that were over 10 years in the making that this committee said, "You must get done in 30 days" in 1 year. So, we are focusing on the ones that Congress has given us deadlines for, and prioritizing safety first.

Mr. MOLINARO. Dear God, I can only imagine that, if 10 years is priority, I can't imagine what 30 is. It is a bit of an embarrassment, wouldn't you think? I mean——

Mr. BROWN [interrupting]. In the last administration—I just went to revisit that exact same hearing, and you got the exact same questions from both sides of the aisle: Why does it take so long?

And the amount of things that we oversee—natural gas, for example—triple what it was 5 years ago that we oversee. Carbon dioxide, hydrogen, $100 billion in incentives. We have zero people, zero full-time employees focused on that. Our ability to get things done is directly proportional to the resources that Congress gives us.

Mr. MOLINARO. Yes, I tend to question that, as a former government official, local government official. I tend to think that it is just—it is a lack of focus and focus on efficient outcomes. I have never been involved in government where somebody didn't say, "What I need is more people to accomplish more," and when we do, they don't accomplish it any quicker, we just have more people doing the same thing. That is my observation. I will be glad to engage in that a little bit further as we get further down the—sorry for the pun—pipeline.

But I want to return, Mr. Grubb. So, the PIPES Act of 2016 convenes this voluntary information-sharing system, a working group. It was recommended that Congress authorize such a program to share pipeline safety information, while providing confidentiality for disclosures.

Could you speak briefly to the benefits of that? And then how might Congress or PHMSA address the concerns about sharing proprietary information?

Mr. GRUBB. Thank you for the question. So, INGAA and its member companies strongly support sharing of information across the sector. It is something that we have done for many years. It is something we will continue to do. We learn from those. We modify our procedures associated with the learnings from other companies, as well as our own. We highly value it. It is something that we share not only with industry sectors, but across the regulatory groups, the governmental groups, the public safety organizations. It is very beneficial. It is something we champion.

With respect to this particular program, I don't have a lot of details on the inner workings of it, and how it would work. I would certainly be willing to look at that and take that back and respond in writing. But I don't have any particular information relative to the inner workings of that information-sharing that you referenced.

Mr. MOLINARO. OK. And, of course, protecting proprietary information, obviously, a priority for the industry. And as we address that, obviously, some guidance and input would be useful.

Mr. Brown, I know that you have already addressed this for my sake, and I am just getting used to the concept of asking questions that have already been answered as a new Member of Congress, but I am happy to continue in that fine tradition.

I wanted to talk, if you could, in my last 30 seconds, the technology pilot program allows operators to deploy that technology that we know will both spur innovation, has environmental benefit, and, of course, increased safety. Can you just reiterate why we have had no participants?

Mr. BROWN. I think Mr. Black has reiterated it a few times, from the perspective of those who are looking to participate. The burden is too high for them to meet. And so, we are happy to work with folks, because we do want to get to the same place, which is using innovative technologies. We don't want to sacrifice safety, and we don't want to sacrifice environmental impacts. But if we can come up with rules of the road there, I mean, that is what we publicly asked for, and we just haven't gotten the suggestion of how to do that.

We have gotten the suggestion that we shouldn't consider the environmental impacts, but the Congress said we have to, is our legal interpretation of that. So, I don't know if I can add too much more, but I do think it is an important point, and I am glad to work with you.

Mr. MOLINARO. We would like that.

And Mr. Chairman, with the degree of priority—without being snide—perhaps we could achieve that in less than 20 years.

Mr. NEHLS. Thank you. The gentleman's time is expired. I now recognize Mr. Huffman for 5 minutes.

Mr. HUFFMAN. Thank you, Mr. Chairman, and thank you for the witnesses.

If I have learned anything in my 10 years in Congress, it is that my colleagues across the aisle pretty much look at any situation, almost every hearing, every bill, as an opportunity to push the fossil fuel agenda. And so, I guess it is no surprise that this week, apparently, the GOP messaging memo was to use the toxic rail tragedy in East Palestine as a pivot to make the case for more fossil fuel pipeline infrastructure. Never mind that that was vinyl chloride, and the pipelines that they want are oil and gas.

I understand why they want to pivot to fossil fuel pipelines, but let's not pretend that that is safe. The truth, if we are honest about it, is that fossil fuel pipelines are disasters waiting to happen, plain and simple.

A great example of that is the Keystone XL, or the Keystone Pipeline. Over the last 12 years, it has had 22 oilspills, including 2 large spills between 2017 and 2019. The GAO issued a report in 2021 regarding this pipeline's safety record, showing that it had been deteriorating, identifying construction issues as the primary contributor. But maybe the more shocking part of that report was that the Keystone Pipeline's terrible safety record was about the same, on average, as other fossil fuel pipelines all over this country.

So, when you put that together with Mr. Caram's testimony showing the number of significant incidents has slightly increased since 2010, I think we have just got to be very clear-eyed about pipeline safety and fossil fuel pipeline safety, specifically, in this country.

On December 22, 2022, Keystone sprang another big leak. This time, 14,000 barrels spilled in Nebraska. It was the pipeline's largest spill since it began operating in 2010, the largest onshore oil-spill in the U.S. since 2013. And PHMSA officials had actually told the GAO way back in 2001 that they had learned lessons, and that they were applying those lessons to inspecting pipelines that were under construction. But here we are.

And so, Mr. Brown, I have been impressed by your knowledge and your competency. You have given great answers to all the questions. If we have to have this spaghetti web of pipelines moving dangerous, toxic, volatile fossil fuel all over this country every day, I want a competent person like you overseeing it, I really do. And I respect the work that you do.

But you also said—you gave us this lessons learned line that we have heard many times before. And I just want to ask if you can appreciate the fact that we have heard this many, many times before. So, as you undertake your rulemaking under the PIPES Act, can you give us something more specific about specific lessons that you are going to put into practice to make these pipelines more safe?

Mr. BROWN. Yes, absolutely. But I just want to get to that really important point that people find it very disconcerting that we focus on lessons

learned. That is, like, by design. That is Congress telling us, and the statutes—
—

Mr. HUFFMAN [interrupting]. You should do it. I am not saying you shouldn't do it.

Mr. BROWN. Yes. No, no, of course. And I just want everybody in the room to understand.

We just had a hearing in the DC Circuit on a rule, and a little bit of the commentary was that unless you can point to a failure specifically, then you can't do a rulemaking. You are not going to survive the legal standard that has been set up by Congress.

Mr. HUFFMAN. Understood. And reclaiming my time, because I want you to learn those lessons and to apply them to your rulemakings, I do want to ask Mr. Caram about hydrogen.

One of the latest things that we are hearing is that all of these new fossil fuel pipelines will soon be carrying hydrogen. We can start blending it right away, and eventually it will be all hydrogen. But it is not that simple, is it, Mr. Caram? We know that the California PUC has just done a study suggesting that there are serious risks and gaps in research and regulations regarding blending hydrogen into fossil fuel pipelines, and that we could very well face leaks from fittings, gaskets, valves, stems.

What are some of the things we should be concerned about before we get too far down the path of hydrogen pipelines?

Mr. CARAM. Yes, thank you for the question. And the Pipeline Safety Trust commissioned a report on the safety of hydrogen pipelines that I encourage everyone to review.

We found that hydrogen has some unique physical characteristics that are very different from methane that make it more dangerous from methane, primarily a much larger flammability range, meaning it more likely to ignite at different concentrations, and it also poses known integrity issues to pipes and components, embrittlement and things like that, which would lead to more failures.

There are some really large knowledge gaps that both PHMSA and the industry, through research organizations, are attempting to fill, but we are years away from answering those questions.

Mr. HUFFMAN. Thank you. And Mr. Chair, I see that I am out of time. I don't have enough time to get into CO2 pipelines, which are not a climate solution. They are for enhanced oil recovery. But we will do that another day. I appreciate your indulgence.

Mr. NEHLS. Thank you, Mr. Huffman. I now yield 5 minutes to Mr. Kean.

Mr. KEAN OF NEW JERSEY. Thank you, Mr. Chairman. And I want to thank all the witnesses for being here today.

As we begin our work to reauthorize pipeline safety legislation, I look forward to working with every single member of the subcommittee to improve pipeline safety through a balanced regulatory approach.

As we have heard, the United States has the largest network of energy pipelines in the world. Oversight of the Pipeline and Hazardous Materials Safety Administration safety programs should be a top priority for this subcommittee in achieving our common goal and our shared goal of zero pipeline incidents.

Mr. Black, the PIPES Act of 2020 creates a technology pilot program that allows operators to test innovative technologies and practices that improve safety in the field. How do operators use innovative technology and practices to improve the pipeline safety of their systems?

And what benefits would a technology pilot program give to operators?

Mr. BLACK. We are excited about the benefits of innovation in inspection and analytic technologies, and we want them to be rolled into how PHMSA regulates pipelines. And I know innovation is very important to you in your part of New Jersey.

Pipeline inspections occur through in-line inspections, so-called smart pig, that travel inside of the pipe. And through the decades, the sensors that have been put on the pipe have gotten more and more sophisticated. We can see more things.

Similarly, back in the office, we are using newer techniques to harness the data from that. But the regulations that PHMSA uses for the schedule that pipelines must address incidents on are more than 20 years old, but yet we have had all these innovations.

So, we are grateful to Congress for creating a pipeline safety demonstration program. Sadly, it hasn't worked. Nobody has applied, because it was made too hard, it went beyond what Congress was suggesting. With some reforms that I am happy to go into, we can road test. PHMSA can see the benefits of new inspection technologies, and how reorienting the schedule that operators need to address risks on improves pipeline safety. That is our goal.

Let's use innovation, new technology, even machine learning to improve. We are doing it, but we need the PHMSA schedules to improve.

Mr. KEAN OF NEW JERSEY. OK. Thank you.

Deputy Administrator Brown, PHMSA has only held one GPAC meeting since January of 2021, and that was on October 20th and 21st of 2021. In the past, you have had many GPAC meetings, while also issuing complicated notice of rulemakings and final rules. Why have you had so few meetings?

Mr. BROWN. So, in the past, we have also had years where we didn't get any rulemakings done. So, we might have had meetings but not done rulemakings.

These meetings are exclusively—at least from my perspective, and all the demands on getting rulemakings done—to advance rulemakings. We have to have a meeting in order for a rulemaking to advance. So—and we want—we had the busiest year in our history in producing pipeline safety rulemakings in 2021, nearly the busiest in 2022.

We have no aversion to having these meetings. We just want to make them as efficient as possible, given the demands we have for all the rulemakings we have to complete.

Mr. KEAN OF NEW JERSEY. OK. Thank you. And Mr. Grubb, can you explain, please explain to this subcommittee, what your committee does to ensure the safety of its pipeline system?

Mr. GRUBB. So, thank you for the question.

So, over time, pipeline operators have invested highly in advancements in pipeline in-line inspection technology. That has improved immensely over time.

The other things are implementation of pipeline safety management systems. We have one at Kinder Morgan. We call it our operations management system. It is really a system that defines our culture.

And there are three main tenets of our operations management system: the first is, every employee has a role in the OMS; the second is instilling a culture of continuous improvement and operational excellence; and the last, and certainly not least, is risk reduction: risk reduction to our assets, to the people that live near our pipelines, to our employees, to our contractors that work in our pipelines. That is a main part of what we do to reduce risk and manage the safety of our pipelines.

The third and last thing is joint industry collaboration, over which GPAC meetings are also crucial and key, interacting with other industries or other industry subject matter experts, Government experts, as well as public safety organization experts.

Mr. KEAN OF NEW JERSEY. Thank you to the entire panel for your testimony.

I will yield back the remainder of my time.

Mr. NEHLS. Thank you. I now yield 5 minutes to Mr. LaMalfa.

Mr. LAMALFA. Thank you, Mr. Chairman. I am glad I was able to slide into the newer, larger second base just in time. Apologies. Overlapping committees make it tough around here, but I appreciate your folks' work with this committee and extremely important topics here on pipelines, which we rely heavily upon in this country, and we need more of. So, let me launch into it here real quickly.

Now, safety is a cornerstone. And that is what much of the work of Congress is seeking to do in increasing safety in all aspects of our lives, including pipelines.

What I note is that the average U.S. company pays approximately $10,000 per employee per year to comply with Federal regulations. So, laws passed with good intent can be completely changed by the time the implementation is completed, or years and decades go by. We see that a lot with the creep of responsibility and jurisdiction.

So, the more input that is received, the rules that the companies have to comply with become less flexible and more burdensome—input by Federal Government and agencies. And the estimate of the National Association of Manufacturers puts the total economic cost of regulations every year at $2 trillion.

In the 2020 PIPES Act, Congress recognized this and sought partially to address it by telling PHMSA to immediately carry out a pilot program to let pipeline operators evaluate new technology and operational changes to improve pipeline safety. But in a way, it is inefficient and also uneconomical.

Again, apologies if some of this is redundant from earlier work today, but since it took too long to send the guidance to operators, the report that they were required to send to Congress a year later was completely void of recommendations, is my understanding.

So, tossing this to Mr. Grubb here, do you know of any recent innovations with technology or operational changes that could have improved pipeline safety under this pilot program?

Mr. GRUBB. I am sorry. Could you repeat the last part of the question?

Mr. LAMALFA. Some of the innovations with technology or operational changes that could have improved the pipeline safety under the pilot program that was to be done under PHMSA.

Mr. GRUBB. Yes, so, to my knowledge, there hasn't been any takers on the technology pilot program because of the complications that Mr. Black talked about.

But I can tell you that pipeline operators are advancing technology on their own.

Mr. LAMALFA. On their own, yes.

Mr. GRUBB. And I can tell you that Kinder Morgan, one of our pipelines was the first pipeline to implement a tool that detects cracks, environmental cracking, in its pipelines. And that was done many, many years ago. It received PHMSA approval. It was the first pipeline to have received that approval. And it has become the industry standard today for detecting cracks.

Mr. LAMALFA. So, the PIPES Act did not really aid in expediting in this in a pilot program?

Mr. GRUBB. No, because of the issues that Mr. Black referenced earlier.

Mr. BLACK. The PIPES Act helped. PHMSA implementation meant nobody took it up. PIPES Act was good. We would like you to continue. Maybe give more direction to PHMSA, have a longer program. Congress was in the right place.

Mr. LAMALFA. OK. What do we need to do to then speed up or do additional work to make PIPES Act effective and something you would want to access?

Mr. BLACK. Say that decisions on pilot program applications don't need to follow NEPA, have an environmental assessment; have a reasonable standard review that is less than something for a permanent waiver, and not use the permanent waiver process; and give this enough time to work. We believe that it can. The technology is an ultrasound, an MRI, and pipelines can improve safety.

Mr. LAMALFA. So, even though it is a pilot program designed to speed things up, it is still under the guidelines of a long-term permitting process. It hasn't really changed.

Mr. BLACK. It is suffering many of the same problems, yes, sir. Two years later, we haven't made notable progress.

Mr. LAMALFA. OK. What do we need to do in order to modify or streamline, massage that act in order to make it so it would be accessible and actually a helpful tool?

Mr. BLACK. Thank you. That would be great, and we would like to work with you. Reauthorize it for a long time. Tell PHMSA to stick with the mandates that you have given. Tell them we don't need to do environmental assessments on R&D projects. We don't need to use the special permit process. Let's make this work.

Mr. LAMALFA. R&D shouldn't need environmental review.

Mr. BLACK. Yes, sir.

Mr. LAMALFA. OK. Makes sense.

Thank you, Mr. Chairman. I yield back. I appreciate the time.

Mr. NEHLS. The gentleman yields. Are there any further questions from any members of the subcommittee who have not been recognized?

Seeing none, that concludes our hearing for today.

I would like to thank each one of the witnesses for your testimony. I would like to thank Ranking Member Payne for being here. I look forward to working with you for the next several months.

I ask unanimous consent that the record of today's hearing remain open until such time as our witnesses have provided answers to any questions that may be submitted to them in writing.

Without objection, so ordered.

I also ask unanimous consent that the record remain open for 15 days for any additional comments or information submitted by Members or witnesses to be included in the record of today's hearing.

Without objection, so ordered.

The subcommittee stands adjourned.

[Whereupon, at 12:10 p.m., the subcommittee was adjourned.]

Submissions for the Record

Letter of March 22, 2023, to Hon. Sam Graves, Chairman, and Hon. Rick Larsen, Ranking Member, Committee on Transportation and Infrastructure, from Geoff Moody, Senior Vice President, Government Relations and Policy, American Fuel and Petrochemical Manufacturers, Submitted for the Record by Hon. Troy E. Nehls

MARCH 22, 2023.

The Honorable SAM GRAVES,
Chairman,
Committee on Transportation and Infrastructure, 2165 Rayburn House Office Building, Washington, DC 20515.

The Honorable RICK LARSEN,
Ranking Member,

Committee on Transportation and Infrastructure, 2164 Rayburn House Office Building, Washington, DC 20515.

RE: Statement for the Record—Pipeline Safety: Reviewing Implementation of the PIPES Act of 2020 and Examining Future Safety Needs

Dear Chairman GRAVES and Ranking Member LARSEN:

The American Fuel & Petrochemical Manufacturers (AFPM) is the leading trade association representing the makers of the fuels that keep us moving, the petro-chemicals that are the essential building blocks for the American manufacturing economy, and the midstream companies that get our feedstocks and products where they need to go. AFPM members make modern life possible and keep America moving and growing as they meet the needs of our nation and local communities, strengthen economic and national security, and support over three million American jobs. AFPM appreciates the opportunity to provide its perspectives on priorities for pipeline safety reauthorization.

As Congress crafts the pipeline safety reauthorization, it should ensure PHMSA focuses on risk-based and cost-effective regulations, promote technology development, and ensure existing programs are working as intended. AFPM recommends Congress consider the following concepts for inclusion in any reauthorization.

Enhance Pipeline Safety Through Data-Driven and Risk Based Initiatives

Congress can help improve safety by encouraging sensible data- and risk-based regulations in the following areas.

- Voluntary information sharing (VIS) can benefit PHMSA's safety mission and AFPM supports a VIS regime provided appropriate protections are in place to ensure such information is used solely as a mechanism to improve pipeline safety. Protecting voluntarily shared information, as well as security sensitive information, must remain a commitment for PHMSA.
- PHMSA standards regulate the inspection of petroleum storage tanks on fixed structures; however, these inspection requirements are rigid and not risk-based, which can lead to unneeded inspections, environmental impacts, and increased risk for workers. Congress

should instruct PHMSA to review current storage tank inspection procedures and implement a risk-based approach that prioritizes worker and environmental safety.

- The National Response Center reporting process currently requires tedious telephonic reporting and PHMSA's update to the incident reporting threshold for liquid pipelines has been delayed. Congress should simplify the National Response Center reporting process and instruct PHMSA to periodically adjust the incident reporting threshold for inflation.

Enable Technology Advancements Through Modernized Pipeline Safety Standards

The regulatory process often lags behind the pace of technological advancements. Congress can foster innovative technologies and support programs that will improve safety standards in the following areas.

- In the PIPES Act of 2021, Congress instructed PHMSA to implement a pilot program to enable operators to use new advanced technologies without major regulations. Unfortunately, the program has not been utilized effectively, as many of the parameters are overly restrictive and have discouraged pipeline operators from the program. Congress should update the Pipeline Safety Technology Demonstration program to encourage operators to utilize this option and deploy advanced technologies under controlled conditions.
- While current federal pipeline safety standards already regulate the design, construction, operation, maintenance, and emergency response for CO2 and Hydrogen pipelines, the regulations could be updated to ensure they are consistent and current. PHMSA should ensure the regulations address any specific safety concerns for this material, but at the same time not be so burdensome that they stifle development. Congress should instruct PHMSA to review and update where necessary specific CO2 and Hydrogen pipeline safety standards.

Improve PHMSA Safety and Enforcement Programs

Congress can provide clarity around PHMSA's safety and enforcement programs in the following areas.

- The rise of interference directed at both pipeline construction and operations creates unsafe conditions and places lives at risk. Pipelines must be protected, and deterrents must be in place to dissuade such actions. PHMSA's safety mission should include authority that ensures pipeline construction and operations remain safe at all times and protected from unsafe conditions and damage.
- In 2021 Congress authorized PHMSA to create a program that would monitor and mitigate methane emissions from natural gas systems. PHMSA expanded the scope of the program beyond natural gas pipelines to include liquid pipelines. AFPM members believe that the inclusion of liquid pipelines has not resulted in any meaningful reductions in hazardous leaks and being overly burdensome on liquid pipeline operators. Congress should clarify that this program does NOT apply to liquid pipelines and refocus the program on natural gas pipelines.
- PHMSA regulations include criteria and procedures for determining the effectiveness of State pipeline excavation damage prevention programs. While there has been significant improvement over the years in avoiding these unintentional hits to pipeline systems, the need to make continuous improvements in avoiding such accidents remains. Congress should continue to encourage maximum participation in these State damage prevention programs by discouraging exemptions from such programs via grant funding to the States.

PHMSA is a small agency with a hugely important mission. While previous reauthorizations authorized PHMSA with resources, many safety priorities have been delayed. AFPM supports Congress reauthorizing PHMSA with the resources to achieve their safety mission. Thank you again for your attention and work on Pipeline Safety reauthorization. We appreciate your leadership on this important issue and look forward to working with lawmakers as the reauthorization process moves forward.

Sincerely,
GEOFF MOODY,
Senior Vice President, Government Relations and Policy,
American Fuel & Petrochemical Manufacturers.

Appendix

Questions from Hon. Troy E. Nehls to Tristan Brown, Deputy Administrator, Pipeline and Hazardous Materials Safety Administration (PHMSA)

Question 1. Is it true the PHMSA has a new Environmental Policy and Justice Division?

ANSWER. There is no formally established Environmental Policy and Justice Division. PHMSA currently has two full time employees with engineering and environmental backgrounds within PHMSA's Office of Planning and Analytics who work on environmental policy. This work has been performed in PHMSA for decades, on an ad hoc basis. PHMSA is considering ways to increase efficiency and reduce costs, and centralize efforts—aimed at tackling environmental issues proactively to help increase the speed of delivery and avoid costly litigation.

Regarding this office:

Question 1.a. Where in the organization is this office located and what are the responsibilities of the personnel in this office?

ANSWER. The effort to create more capacity for proactive analysis, as previously mentioned, is ongoing.

Question 1.b. What is its annual budget?

ANSWER. There is no formally established Environmental Policy and Justice Division so it does not have a budget.

Question 1.c. Is this office involved in the development of rulemakings?

ANSWER. PHMSA currently has two full time employees focused on ensuring the agency meets the requirements established by Congress for NEPA reviews, including for rulemakings.

Question 1.d. Can you please list the number of personnel and corresponding title and GS-level for each personnel located in this office?

ANSWER. There are currently three personnel who work on environmental policy, all with engineering (and environmental) backgrounds:

- GS–0028–15 Supervisory Environmental Protection Specialist

- GS–0028–14 Lead Environmental Protection Specialist
- GS–0028–09/11/12 Environmental Protection Specialist

Question 1.e. Are there additional positions to be filled in this Division and if so, can you please list those vacant positions?

ANSWER. The Office of Planning and Analytics is in the process of hiring a GS– 0028–13/14 Lead Environmental Protection Specialist to conduct state, local, tribal, and territorial reviews to ensure that PHMSA is meeting its statutory requirements under Section 106 of the National Historic Preservation Act. The office does not have any plans to hire any additional environmental protection specialists other than the ones previously mentioned.

Question 2. We've learned that PHMSA has six pipeline safety climate change expert positions that are open. Can you please detail the job titles, GS-level, position duties, and where they are located in PHMSA's organizational structure? How many of these positions have been filled?

ANSWER. The FY 2023 appropriations provided funding for six (6) positions to support the requirements in Executive Order 14008 (Tackling the Climate Crisis at Home and Abroad), among several other duties consistent with their respective job series. Four positions have been filled.

The positions are as follows:

- Office of Planning and Analytics (4 positions)
    - GS–0028–15 Supervisory Environmental Protection Specialist (Filled)
    - GS–0028–9/11/12/13/14 Environmental Protection Specialist (Vacant)
    - GS–0028–9/11/12/13/14 Environmental Protection Specialist (Vacant)
    - GS–0028–9/11/12 Environmental Protection Specialist (Filled)
- Office of Pipeline Safety (OPS, 2 positions)
- GS–2101–14 Transportation Specialist, Standards and Rulemaking Division (Filled). This position will help lead the development of rulemaking efforts related to methane emissions reduction for gas transmission, gas gathering, gas distribution, liquefied natural gas (LNG) and hazardous liquid pipeline facilities. This individual will also be involved in the implementation, including stakeholder outreach and engagement, of these efforts to reduce the safety and climate impact of pipeline infrastructure.

- GS–2101–13 Transportation Specialist, Standards and Rulemaking Division (Filled). This position will support the development of rulemaking efforts related to methane emissions reduction for gas transmission, gas gathering, gas distribution, LNG, and hazardous liquid pipeline facilities. This individual will also be involved in the implementation, including stakeholder outreach and engagement, of these efforts to reduce the safety and climate impact of pipeline infrastructure.

Question 3. Section 102(b)(1) of the Protecting Our Infrastructure of Pipelines and Enhancing Safety Act of 2020 (PIPES 2020) mandated PHMSA hire eight full time employees with subject matter in pipeline safety, pipeline facilities, and pipeline systems to finalize outstanding rulemakings and fulfill congressional mandates. PHMSA's FY 2024 budget request states, "In furtherance of its rulemaking efforts and in accordance with the PIPES Act of 2020, PHMSA is in the process of completing its hiring of eight full-time employees to help finalize outstanding rulemakings and fulfill congressional mandates."

Question 3.a. Can you please list the number of positions that have been filled related to this mandate along with their associated titles and GS-level?

ANSWER. PHMSA has filled all eight (8) positions:

- GS–0801–14 General Engineer
- GS–0801–14 General Engineer
- GS–0801–13 General Engineer
- GS–0905–14 Attorney Advisor
- GS–1515–13 Operations Research Analyst
- GS–0110–13 Economist
- GS–1083–13 Technical Writer
- GS–2101–9 Transportation Specialist

Question 3.b. Can you please also share what subject matter expertise in pipeline safety, pipeline facilities, and pipeline systems each person hired has?

ANSWER. Almost all of the individuals hired to fill the above positions had experience in pipeline safety or the pipeline industry. The attorney and economist had transferable experience to PHMSA's regulatory program, and they have been instrumental to the progress that PHMSA has made in the last two years in one of the busiest rulemaking periods in the agency's history.

Question 3.c. Can you describe the important contributions to pipeline safety that each person has made since onboarding?

ANSWER. Over the last two years, PHMSA has achieved significant milestones in rulemaking, supported by personnel with pipeline safety rulemaking responsibility, including the eight (8) hired thus far under this mandate. Significant contributions include development of Notices of Proposed Rulemakings (NPRMs) for Gas Pipeline Leak Detection, Safety of Gas Distribution Pipelines, safety of Carbon Dioxide Pipelines, Hazardous Liquid Repair Criteria, LNG Safety, and Pipeline Operational Status; information collection requests from OMB for the recently published Valve Rule; development of congressionally mandated studies and reports; and performance of regulatory impact analysis and regulatory flexibility analysis for rulemakings.

The eight (8) personnel hired under this mandate have the following responsibilities:

- GS–0801–14 General Engineer, GS–0801–14 General Engineer, and GS–0801– 13 General Engineer: provide technical expertise and engineering and science-based analysis in support of rulemaking, improvement of reliability and serviceability of the pipeline transportation network, and development of studies and reports mandated in the PIPES Act of 2020.
- GS–2101–9 Transportation Specialist: responsible for specializing in rule-making, perform research and analysis related to the development of regulatory changes and interpretation of regulations, participate in the development of proposed and final rulemaking documents, including environmental and economic impact statements, evaluation of public comments, and incorporation of legal input on proposed regulatory changes.
- GS–0905–14 Attorney Advisor: responsible for providing legal advice in support of the policy development efforts, development of rulemakings and implementing guidance, and defense of the same from administrative and appellate litigation. The incumbent ensures rulemakings developed are evidence-based, legally sound, and otherwise satisfy the myriad of procedural and substantive requirements under Federal administrative law and PHMSA's safety statutes.
- GS–1515–13 Operations Research Analyst: responsible for providing data analysis and visualization and reports to various PHMSA offices

to address inquiries, respond to mandate and audit activities, and support Federal regulation compliance.

- GS–0110–13 Economist: responsible for conducting economic research to understand economic and industry trends that influence risks to pipelines, the transportation of hazardous materials, and/or related industries. The incumbent performs regulatory impact analysis and regulatory flexibility analysis, assessing the costs, benefits, efficiency, and impacts of PHMSA's regulatory and safety programs.
- GS–1083–13 Technical Writer: responsible for providing writing and editorial support for rulemaking and to respond to Congressional reports, mandates, and other requirements.

Question 3.d. Can you please list the number of vacant positions, along with their associated titles and GS-level?

ANSWER. All eight (8) positions have been filled.

Question 4. Section 102(b)(2) of PIPES 2020 mandated that PHMSA ensure the number of full-time positions for the pipeline inspection and enforcement personnel in the Office of Pipeline Safety does not fall below 224 for fiscal year 2021, 235 for fiscal year 2022, and 247 for fiscal year 2023.

Question 4.a. Can you please let us know the current number of full-time pipeline inspection and enforcement personnel employed by PHMSA?

ANSWER. In compliance with this requirement, as of May 18, 2023, PHMSA has 247 full-time pipeline safety inspection and enforcement positions. We have continued to accept applications and recruit for these positions, and recently we received approval and applied additional retention and recruitment incentives. As of September 30, 2023, 214 of these positions have been filled.

Question 4.b. According to your testimony, "PHMSA has kept up with the PIPES Act of 2020 hiring mandates—both for inspectors as well as for regulatory personnel. . ." However, according to information provided by PHMSA to staff, PHMSA currently only has 207 pipeline inspection and enforcement personnel. If this number is accurate, do you believe it is compliant with the number set out in Section 102? ANSWER. The 2020 PIPES Act requires PHMSA to ensure that the number of full-time positions for pipeline inspection and enforcement (I&E) personnel in the Office of Pipeline Safety of the Administration does not fall below 224 for fiscal year 2021, 235 for fiscal year 2022 and 247 for fiscal year 2023. PHMSA has established the openings and continued to receive applications consistent with the requirement, and as noted above has filled the regulatory personnel positions; further:

- With receipt of appropriations in the FY22 cycle, which included funding for additional I&E positions in December of FY22, PHMSA has put a robust and aggressive recruitment and retention strategy in place, focusing on soon-to-be college graduates and recent graduates for entry level opportunities, and internships that will create a pipeline of future engineers. PHMSA is also working on providing student loan repayment to support its recruitment and retention efforts.
- PHMSA successfully onboarded 13 full-time I&E employees in FY 2021, 24 I&E employees in FY 2022, and 24 I&E employees to date in FY 2023. However, more staff separated from PHMSA's Office of Pipeline Safety between fiscal years 2021 and 2022; in FY 2023, 14 I&E personnel separated from PHMSA. As Congress has recognized, these are also hard to fill positions—PHMSA competes with the private sector to attract and retain qualified pipeline inspectors and engineering analysts. The commercial energy industry and at least one other federal agency offer candidates higher salaries and provide incentives. That is why PHMSA requested and received approval for a special rate table from OPM to provide a 35 percent premium to grade 5–12 engineer inspectors in its five regional offices, in line with other Federal agencies (on account of Congress' directive and the fierce competition from industry salaries), which was instituted as of May of 2023. PHMSA also continues to implement other hiring solutions, such as recruitment and retention incentives, tuition assistance, and student loan repayment.
- The job pool is small because the work is technically challenging—requiring engineering and other technical degrees—and requires individuals who are willing to travel frequently and work in harsh outdoor and sometimes hazardous conditions.
- Like any organization, public or private sector, large or small, PHMSA will have vacancies due to departures and retirements. In the wake of the COVID- 19 pandemic, some employees are choosing to pursue work that does not involve frequent travel across the country. While PHMSA has been successful in bringing many new inspectors and investigators onboard, it has also lost a similar number to retirement and resignation. That is the reason that the number of positions does not equal the number of current staff onboard. We have

noticed similar challenges for state pipeline inspector openings as well.

Question 5. Section 102 (c) of PIPES 2020 provided PHMSA with the authority to use incentives to retain a qualified workforce, including inspection and enforce- ment personnel, attorneys, and subject matter experts at the Office of Pipeline Safety at PHMSA. Can you please provide a status update on the implementation of the following incentives and if they have been utilized, including the number of the times they have been used:

Question 5.a. Special pay rates permitted under section 5305 of title 5, United States Code;

ANSWER. In 2021, PHMSA engaged an outside contractor to determine the parameters for increasing its special pay rates—including for example the geographic areas and pay grades—necessary for OPM approval of the special pay rates. In FY22, PHMSA received additional funding to provide financial incentives such as increasing the pay rates. In August of 2022, PHMSA/DOT submitted to the Office of Personnel Management (OPM) for approval an increase in the special pay rates for engineer/inspectors. PHMSA received final approval from OPM on April 17, 2023, and promptly implemented the new pay rates.

Question 5.b. Repayment of student loans permitted under section 5379 of title 5, United States Code;

ANSWER. PHMSA has developed a plan to use the repayment of student loans as a retention incentive for its 0800 engineering series employees. PHMSA is working to develop an internal process for applications of the repayment and expects to implement this retention incentive in FY23.

Question 5.c. Tuition assistance permitted under chapter 41 of title 5, United States Code;

ANSWER. PHMSA is in the process of creating a new tuition assistance program.

Question 5.d. Recruitment incentives permitted under section 5753 of title 5, United States Code; and

ANSWER. PHMSA actively utilizes recruitment incentives for highly qualified applicants for mission-critical occupations.

Question 5.e. Retention incentives permitted under section 5754 of title 5, United States Code.

ANSWER. PHMSA is examining ways to offer group retention incentives to combat increased turnover experienced last year. The group retention incentives could be offered to eligible inspectors and engineers in positions that are hard to fill or have high turnover. The incentives would not exceed 10% of the proposed salary per recipient and would help PHMSA compete with industry salaries. The incentive could be coupled with a continuation of service agreement as called for in the PIPES Act. PHMSA has strongly emphasized recruitment efforts and has successfully implemented strategies to attract and retain qualified talent. However, based on recent data, PHMSA recognizes that retention has become a growing concern across the sector, and as such, we are dedicating additional resources to retention efforts moving forward. We believe that these incentives will be effective in our efforts to retain our talented workforce. Once implemented, we will track and measure their effectiveness to ensure they have the desired impact.

Question 6. Are there any recruitment and retention incentives not currently held by PHMSA or included in PIPES 2020 that you believe would increase PHMSA's ability to recruit a qualified workforce? If so, can you please detail them as well as any needed congressional authority to be able to utilize them?

ANSWER. PHMSA has recently onboarded a new director for human resources who is currently reviewing every available incentive possible and proceeding with implementation for those most feasible for PHMSA and our incumbents. There is an approval process by which agencies like PHMSA may utilize existing recruitment and retention incentives—i.e., the authorities PHMSA has are not self-executing. Additionally, PHMSA invests in employee leadership development—including three tiers of leadership development internal programming as well as external coaching services that have been shown to both improve employee morale as well as increase employee productivity. These areas show additional promise for attracting and retaining talent. While PHMSA is not requesting any additional authorities at this time, the agency is continuing to explore opportunities to ensure we are able to recruit and retain a robust pipeline safety workforce.

Question 7. Can you please explain why PHMSA has not held a Gas Pipeline Advisory Committee (GPAC) meeting on the class location rule when Congress required in the 2020 PIPES Act to do so by the end of calendar year 2021? When does PHMSA plan to hold this meeting?

ANSWER. PHMSA has attempted to successfully balance its existing statutory directives with new directives from Congress, many of which have included aggressive deadlines, including this one. PHMSA recognizes that obtaining input from the Gas Pipeline Advisory Committee and Liquid Pipeline Advisory Committee is a crucial part of its rulemaking process, including for the Class Location Change Requirements proposed rule. In 2021, after the passage of the 2020 PIPES Act, the Government Accountability Office completed a report on the Keystone pipeline, noting the unique circumstances under which PHMSA issued and maintained a special permit for that line. In response, PHMSA initiated an independent third-party assessment of its special permit process and the safety conditions required of operators for class location special permits—across the board. To optimize use of the GPAC and ensure that all relevant information related to the proposed rule is available to the GPAC, and to make efficient use of the agency resources and the Committee's time, PHMSA planned to hold a GPAC meeting on the class location rulemaking after the conclusion of the special permit process review. PHMSA recently announced the GPAC meeting will occur the week of November 27, 2023. The meeting will consider both the Class Location Rule as well as the Leak Detection and Repair proposed Rule. Holding a PAC meeting before the independent report would deprive the PAC of that information and could further delay the finalization of the rule. PHMSA has already reached out to GPAC members and scheduled a meeting to review this NPRM, as well as PHMSA's NPRM for Gas Pipeline Leak Detection and Repair.

Question 8. When does PHMSA anticipate issuing a final class location rule?

ANSWER. As stated above, PHMSA has made public an expected date for this meeting, to consider both the Class Location Rule as well as the Leak Detection and Repair Proposed Rule. PHMSA currently anticipates the rule to be published by the end of 2024.

Question 9. Deputy Administrator Brown, in your response to a question from Representative Marc Molinaro (R–NY) about the class location rule, you stated that PHMSA is, " . . . focusing on the [rules] that Congress has given us deadlines for and prioritizing safety first." You also characterized the class location rule as being focused on, " . . . efficiency, potential impacts for reducing environmental emissions. We focus on safety first." The Interstate Natural Gas Association of America (INGAA) believes that finalizing this rule will significantly benefit safety by allowing its members to reallocate between $200–$300 million per year that could be invested in running advanced inspection tools on 25,000 miles of pipe. In addition, INGAA believes that finalizing this rule would stop up to 800 million cubic feet of natural gas from being released to the atmosphere each year. Completing the class location rule would clearly fulfil the broader safety and environmental objectives of PHMSA. Furthermore, the class location rule had a mandate from Congress in the PIPES Act of 2020 to hold a Gas Pipeline Advisory Committee (GPAC) meeting on it by the end of 2021. We have several inquiries on this topic:

Question 9.a. Can you please explain why your agency has yet to prioritize, and possibly ignored, this mandate thus far?

ANSWER. Please see response to question 7 above.

Further, with respect to the characterization of the rule, I would also refer you to the August 1, 2022 letter to me from then Ranking Members Graves and Crawford as well as Senators Wicker and Fischer who noted "Class location change regulations mandate that pipeline operators undertake certain actions when population density changes, such as replacing pipeline segments that result in large releases of gas into the atmosphere. Modernizing these requirements has the potential to create efficiencies for operators and further protect the environment by minimizing these releases."

Question 9.b. Can you please explain why your agency has yet to prioritize the implementation of the class location rule when there are both safety and environmental benefits?

ANSWER. In addition to the context provided in response to subpart a of your question, since the beginning of the Administration, PHMSA has prioritized Congressional mandates (and associated) deadlines with ensuring PHMSA advances the rules that can have the greatest impact on safety. PHMSA in turn prioritized finalizing three rules that were twelve years in the making—spanning four Presidential administrations—which stemmed from tragic incidents involving pipeline-related fatalities such as the 2010 PG&E San Bruno, CA incident.

Question 9.c. When specifically does PHMSA plan to hold a GPAC meeting on the class location Notice of Proposed Rulemaking (NPRM)?

ANSWER. PHMSA has already reached out to GPAC members to schedule a meeting to review this NPRM, as well as PHMSA's NPRM for Gas Pipeline Leak Detection and Repair in the Fall of 2023.

Question 9.d. Why did PHMSA hold its only GPAC meeting since the start of 2021 on the Periodic Updates of Regulatory References to Technical Standards and Miscellaneous Amendments NPRM, which was not Congressionally mandated?

ANSWER. Please see response to question 7 above.

Additionally, periodic Updates of Regulatory References to Technical Standards and Miscellaneous Amendments proposed rule fulfills a Congressional mandate. 49U.S.C. § 60102(l), Updating Standards, states that the Secretary shall update incorporated industry standards that have been adopted as part of the Federal pipeline safety regulatory program. The

"Periodic Updates of Regulatory References to Technical Standards and Miscellaneous Amendments" NPRM is a rulemaking that addresses 26 such standards—more than 30 percent of the standards that are incorporated by reference in the pipeline safety regulations. As required by 49 U.S.C. § 60115, PHMSA convened a GPAC meeting on this rule.

On March 8, I testified "As required by Congress, PHMSA continues to update the Federal pipeline safety regulations (PSRs) to reflect new and revised voluntary consensus standards developed and adopted by standards-setting bodies (see e.g., PHMSA's periodic standards update rulemakings). We understand how important updating and aligning standards can be to ensure the PSR include up-to-date standards that reflect current best practices and technologies—and to serve as a higher bar, from which the regulated community can continue to improve."

Question 9.e. Can you explain what process your agency has for processing Congressional rulemaking mandates?

ANSWER. Generally, when PHMSA receives a congressional mandate to complete a rulemaking, PHMSA works with personnel in both the program office and the Chief Counsel's office to initiate the rulemaking by requesting a Regulatory Identification Number (RIN)—which is then submitted to the Office of the Secretary's Office of General Counsel for approval and inclusion in the Unified Agenda. PHMSA staff then reviews the statutory language in the mandate and begins drafting proposed regulations and a Notice of Proposed Regulations (NPRM). Once drafted, PHMSA will then issue the NPRM in the Federal Register for public comment. If required, PHMSA will hold a meeting of the relevant Pipeline Advisory Committee to review and discuss the statutory language and any proposed regulatory text, as well as any comments received. PHMSA will then issue a final rule adopting new or updated regulations. Occasionally, in response to public comments on the NPRM, PHMSA may elect to issue a Supplemental NPRM if we need to adjust the proposed regulations or ask additional questions before finalizing the regulations.

As part of this process, PHMSA continuously works to balance completion of outstanding congressional mandates for rulemaking, as well as other rulemakings initiated as a result of NTSB recommendations, concerns arising from incidents, and other sources. As noted at the hearing, the agency—through Republican and Democratic administrations, since its inception—has typically had more congressional mandates than it has resources to complete under the timeframes provided. PHMSA in turn, attempts to triage mandates based on the overlapping congressional mandates and with safety being the top priority.

Question 10. Can you please detail the number and dates of meetings PHMSA has held of the GPAC and Liquid Pipeline Advisory Committee (LPAC) since January 1, 2021? What is the agency's target for annual frequency of meetings for both of these respective committees?

ANSWER. Since January 1, 2021, PHMSA has held three (3) pipeline advisory committee meetings:

- October 20, 2021: The LPAC and GPAC both met jointly to discuss the notice of proposed rulemaking (NPRM) titled: "Periodic Updates of Regulatory References to Technical Standards and Miscellaneous Amendments." The NPRM is relevant to both gas and liquid pipeline safety.
- October 21, 2021: The LPAC and GPAC met jointly to discuss the Standards Update notice of proposed rulemaking.
- August 17, 2022: The LPAC met to discuss the interim final rulemaking (IFR) titled: "Unusually Sensitive Areas for the Great Lakes, Coastal Beaches, and Certain Coastal Waters."

Given the statutory role of the LPAC and GPAC, PHMSA schedules meetings of those committees to review and discuss specific rulemakings. Accordingly, the schedule is based on the progression of PHMSA's regulatory agenda—with the goal of maximizing efficiency of the meetings. PHMSA also meets with most of the organizations represented on the advisory committees on an individual basis—again, for efficiency.

Question 11. What do you believe is the appropriate role for the GPAC and LPAC? What value does the GPAC and LPAC provide to PHMSA in the rulemaking process?

ANSWER. Under 49 U.S.C. § 60115, the Gas Pipeline Advisory Committee (GPAC) and the Liquid Pipeline Advisory Committee (LPAC) are peer-review committees regarding carrying out 49 U.S.C. Ch. 601. As required by law, the GPAC and LPAC, respectively, are composed of a broad spectrum of stakeholders with equities in PHMSA proceedings, with equal representation from government, industry, and the public. These committees offer input that is a statutorily mandated part of PHMSA's rulemaking process. The committees review PHMSA's regulatory initiatives to determine technical feasibility, reasonableness, cost-effectiveness, and practicability.

PHMSA values the feedback from its advisory committees for Office of Pipeline Safety rulemakings. They play a key role in the federal pipeline safety rulemaking process. The input they provide helps PHMSA ensure a constructive, in-depth look at our rules, helps inform potential changes to them, and allows PHMSA to make more effective rules to govern pipeline safety.

Question 12. PHMSA has stated that litigation is impacting its ability to finalize rules. Do entities like the Pipeline Advisory Committees reduce litigation risk?

ANSWER. PHMSA welcomes stakeholder feedback during all stages of its rule-making process. As required by law, PHMSA's Gas Pipeline Advisory Committee and the Liquid Pipeline Advisory Committee advisory committees review PHMSA's regulatory initiatives to determine technical feasibility, reasonableness, cost-effectiveness, and practicability. PHMSA values the feedback from its advisory committees for Office of Pipeline Safety rulemakings and that feedback can help ensure PHMSA's rulemakings are more legally defensible.

Question 13. Congress mandated a 2-year deadline for PHMSA to complete the idled pipeline rulemaking as part of the PIPES 2020; however, PHMSA missed the congressionally established deadline to complete this rulemaking this past December. This rulemaking is important to the regulated community as pipeline operators need regulatory certainty from PHMSA for situations when pipeline operators suspend normal service on pipeline systems due to changes in market conditions in which the status of these pipelines is neither "active" nor "abandoned."

Question 13.a. Why hasn't PHMSA issued a Notice of Proposed Rulemaking (NPRM) addressing idled pipelines when Congress specifically required a rule to be promulgated by the end of December 2022?

ANSWER. It is important to note that rulemaking is designed to be an iterative process that encourages maximum participation by all stakeholders and rigorous analysis in support of decision making. PHMSA uses its limited resources to promulgate comprehensive rules that protect the public and the environment and meet our statutory requirement for rules. The PIPES Act 2020 included an aggressive timeline for the agency to advance rulemakings (as well as other congressional directives), and PHMSA has completed several of these rulemakings and other legacy rulemakings with important safety impacts related to remote-control valves, gas gathering pipelines, and increased protections for unusually sensitive areas. At the same time, as mandated by the PIPES Act of 2020, PHMSA has recently issued proposed rules related to Gas Pipeline Leak Detection, and the Safety of Gas Distribution pipelines. PHMSA anticipates publishing an NPRM on idled pipelines by the first quarter of 2024. After the NPRM is published, PHMSA will schedule and facilitate the necessary GPAC/LPAC meeting and work to complete a final rule.

Question 13.b. When will PHMSA finalize the idled pipe rulemaking?

ANSWER. PHMSA currently anticipates publishing an NPRM on idled pipelines by the first quarter of 2024. After the NPRM is published, PHMSA will schedule and facilitate the necessary GPAC/LPAC meeting and work to complete a final rule.

Question 14. Section 104 of PIPES 2020 authorized PHMSA to establish a Pipeline Safety Enhancement Program (program) as a means for operators to demonstrate new technologies and procedures that advance pipeline safety. A number of pipeline operator stakeholders are encouraging Congress to reauthorize the program as it has not been utilized to date. PHMSA says operators have not applied for the program. Operators tell us they want to apply, but PHMSA has added additional requirements that make participation infeasible.

Question 14.a. Do you believe a program that allows operators to test safety procedures and technology more swiftly is beneficial to improving pipeline safety as a whole?

ANSWER. PHMSA believes that testing new safety procedures and technologies can be beneficial to improving pipeline safety. Because of the potential serious consequences to people and the environment, as well as existing congressional mandates, testing under the current program requires an evaluation process and stakeholder notice and comment prior to in-situ testing. PHMSA has sought comment on ways to improve the program and would like to work with Congress to find ways to improve the program.

Question 14.b. If an operator would like to test a new safety practice or technology in the field, would it take less time to utilize the technology pilot program or acquire a special permit?

ANSWER. The timing would be based upon the specific application. While both follow similar paths and require, in accordance with Federal law, the completion of an environmental assessment, the Section 104 process includes more specific criteria that need to be met in order to apply, such as the location being outside prohibited areas (a high population area (HPA), high consequence area (HCAs) or an unusually sensitive area (USA)).

Question 14.c. Do you have any suggestions for how Congress could improve the program so operators will be more inclined to utilize it?

ANSWER. PHMSA sought public comment on ways to improve the program and would like to work with the Congress to find ways to improve the program.

Question 14.d. If the program were to be extended for a longer period, for example, 10 years or more, how might that impact PHMSA's ability to implement the program and how might it affect its use by operators?

ANSWER. An extension of time would allow for a more thorough evaluation of the potential benefit of the new safety procedures and technologies by the operators and PHMSA that could be applied throughout the industry.

Question 15. The PHMSA program implementation guidance for the Pipeline Safety Enhancement Program (program) requires applicants to meet conditions and follow processes beyond those enacted by Congress. To date, no pipeline operator has applied to PHMSA under these conditions.

Question 15.a. During the March 8, 2023, hearing, Deputy Administrator Brown testified PHMSA believes Federal law requires National Environmental Policy Act (NEPA) analysis before approving a program. What is the statutory or legal basis for this opinion?

ANSWER. PHMSA's consideration and approval of applications for pipeline safety enhancement programs established in Section 104 of the PIPES Act of 2020—is a discretionary Federal agency action or decision, which is presumptively subject to the National Environmental Policy Act (42 U.S.C. 4321 et seq., NEPA), as well as any applicable implementing regulations and policies issued by the agency or CEQ (40 CFR parts 1500–1508). Those legal authorities require consideration of the impacts of those Federal actions or decisions on the human environment.

Question 15.b. In a February 9, 2022, meeting between industry stakeholders and PHMSA career staff, representatives of the PHMSA Office of Chief Counsel agreed the program does not trigger a statutory obligation to perform NEPA analysis. However, Deputy Administrator Brown testified to Congress on March 8, 2023, that PHMSA is obligated by law to conduct NEPA analysis. Why did Deputy Administrator Brown testify to Congress contrary to the legal opinion of the PHMSA Chief Counsel's Office?

ANSWER. Like any discretionary Federal agency action or decision, PHMSA's consideration and approval of applications for pipeline safety enhancement programs established in Section 104 of the PIPES Act of 2020 are presumptively subject to the National Environmental Policy Act (42 U.S.C. 4321 et seq., NEPA), and to applicable implementing regulations and policies issued by the agency or CEQ (40 CFR parts 1500–1508).

Accordingly, during the March 8, 2023, hearing, I testified PHMSA believes Federal law requires National Environmental Policy Act (NEPA) analysis before approving a program. PHMSA's Office of Chief Counsel staff is unaware of any statements inconsistent with the March 8, 2023, testimony, including in any meetings between industry stakeholders and PHMSA staff.

Question 15.c. Does PHMSA consider establishing a testing program as authorized by Sec.104 a major Federal action significantly affecting the quality of the human environment?

ANSWER. PHMSA's consideration and approval of applications for pipeline safety enhancement programs established in Section 104 of the PIPES Act of 2020 is a discretionary Federal agency action or decision, which is presumptively subject to the National Environmental Policy Act (42 U.S.C. 4321 et seq., NEPA), as well as any applicable implementing regulations and policies issued by the agency or CEQ (40 CFR parts 1500–1508). Those legal authorities require consideration of the impacts of those Federal actions or decisions on the human environment.

Question 15.d. If so, representatives of PHMSA's Chief Counsel's Office agreed during a February 9, 2022, meeting with industry stakeholders that the program did not represent a major Federal action that would trigger NEPA analysis. If PHMSA has changed its opinion on this, what is the statutory or legal basis for this change?

ANSWER. PHMSA understands its consideration and approval of applications for pipeline safety enhancement programs established in Section 104 of the PIPES Act of 2020—like any discretionary Federal agency action or decision—are presumptively subject to the National Environmental Policy Act (42 U.S.C. 4321 et seq., NEPA), as well as any applicable implementing regulations and policies issued by the agency or CEQ (40 CFR parts 1500–1508). Those legal authorities require consideration of the impacts of those Federal actions or decisions on the human environment.

Section 104 of the PIPES Act of 2020 contains no exception to NEPA's procedural requirements. Rather, Section 104 explicitly contemplates that PHMSA may, in reviewing applications for pipeline safety enhancement programs, employ its longstanding special permit procedures issued pursuant to 49 U.S.C. 60118(c)(1)(A). See 49 U.S.C. 60142(d)(2)(A). NEPA review is a critical component of those special permit procedures. See 49 CFR 190.341(c).

PHMSA's Office of Chief Counsel staff are unaware of any statements by PHMSA staff that are contrary to the above position or inconsistent with the March 8, 2023, testimony.

Question 15.e. Has PHMSA ever previously required a research and development project, including those demonstrating promising technologies or analytic methods, to conduct a NEPA analysis?

ANSWER. Unlike traditional pipeline safety-related research and development activated, directed, or funded by PHMSA (which are typically performed using some combination of software modeling or lab or institution-based testing under controlled conditions), pipeline safety enhancement programs are expected to be conducted on operational gas and hazardous liquid pipelines. Those programs are also expected to require a waiver of existing PHMSA safety regulations that would otherwise prevent the use of the innovative technologies and operational practices being tested. Therefore, novel technologies and operational practices employed in pipeline safety enhancement programs could involve markedly different potential impacts on the human environment than PHMSA's own traditional pipeline safety-related research and development activity—including positive impacts, which would of course be welcome. In addition, Congress included within that provision limitations on pipeline safety enhancement programs evincing concern with the potential safety and environmental consequences of the novel technologies and operational practices in those programs, such as program duration, mileage limitations, location limitations and safety thresholds.

Question 15.f. What is the statutory or legal basis for PHMSA requiring program applicants to submit applications in accordance with PHMSA's special permit process?

ANSWER. Section 104 explicitly contemplates that PHMSA may, in reviewing applications for pipeline safety enhancement programs, employ its longstanding special permit procedures issued pursuant to 49 U.S.C. 60118(c)(1)(A). See 49 U.S.C. 60142(d)(2)(A). NEPA review is a component of those special permit procedures. See 49 CFR 190.341(c).

Question 16. Section 111 of the PIPES 2020 sets out that PHMSA, subject to the availability of funds appropriated by Congress, may establish a National Center of Excellence for LNG Safety. Per the requirements of Section 111, PHMSA submitted a report to Congress titled National Center of Excellence for Liquefied Natural Gas Safety. One proposal included in the report to Congress is that PHMSA institute an LNG Separate Center of Excellence at a new location, as discussed in the report under "Model 3," with an estimated annual cost of $8.4 million. The Consolidated Appropriations Act, 2023 provided PHSMA $8.4 million for PHMSA to establish a Center of Excellence.

Question 16.a. When will PHMSA issue a decision on if it will utilize its authority in Section 111 of the PIPES 2020 and FY 2023 appropriated funds to create a National Center of Excellence of LNG Safety?

ANSWER. As stated during my testimony in response to Rep. Brian Babin of Texas, PHMSA has already initiated planning for the establishment of the Center. Since then, PHMSA has also hosted a public meeting on the subject (in May 2023), made visits to potential locations for the center, and assigned a senior leader, with considerable LNG experience, to oversee the planning, creation and management of the Center.

Question 16.b. If PHMSA does not intend to pursue Model 3, how will it make up for the shortfalls and gaps in knowledge that would occur?

ANSWER. PHMSA plans to pursue Model 3 to establish the LNG Center of Excellence.

Question 17. How is PHMSA working to improve the Federal permitting process for energy infrastructure projects in instances where PHMSA provides the regulatory expertise on the safe design and operations and FERC holds construction authorization authority, especially for large-scale LNG projects?

ANSWER. On August 31, 2018, PHMSA and FERC entered into a Memorandum of Understanding (MOU) that describes each agency's respective roles and responsibilities concerning the siting, construction, and operation of LNG facilities, and established a new coordination framework to streamline the permitting approval process for those facilities. Specifically, under the terms of the MOU, PHMSA is responsible for reviewing whether an application meets the siting criteria and safety requirements contained in Part 193, Subpart B.

In accordance with the MOU, upon receiving an application from FERC, PHMSA will issue a Letter of Determination (LOD) to FERC with its analysis and conclusions 30 days prior to FERC's final NEPA document issuance. The agencies' coordination has helped improve the environmental reviews on several LNG export terminal applications. To date, PHMSA has issued LODs for 25 projects.

Currently, PHMSA is reviewing five (5) applications and will issue LODs once PHMSA has completed its Part 193, Subpart B review for the following proposed projects:

- Golden Pass Uprate Project (CP20–459–000)
- Calcasieu Pass Uprate Amendment Project (CP22–25–000)
- Venture Global CP2 (CP22–21–000)
- Driftwood LNG (CP17–117–000)

- Plaquemines Uprate (CP17–66–000)

Question 18. Have there been instances where project developers are made to redesign existing facilities to comply with requirements from different Federal inspectors and regulators? If so, does PHMSA have any recommendations or plans to improve this process to avoid such instances?

ANSWER. PHMSA is unaware of project developers being required to redesign existing facilities to comply with other Federal standards. PHMSA performs comprehensive safety inspections of LNG facilities and monitors and enforces an LNG operator's compliance with the DOT's Pipeline Safety regulations. In accordance with § 191.22 (c)(1)(iii), LNG operators are required to notify PHMSA 60 days prior to construction of a new LNG facility. Currently, PHMSA reviews the final design records for compliance with Part 193, Subpart C—Design, while construction activities are in progress.

Question 19. United States LNG exports have grown rapidly and safely and it is important that regulations that govern these facilities are modernized without delay. PIPES 2020 included two provisions that impacted LNG—the creation of risk-based regulations for LNG facilities, and the creation of a National Center of Excellence for LNG safety. Can you give us an update on where PHMSA stands, and if PHMSA will meet the timeline set forth by the PIPES 2020?

ANSWER. With respect to regulations for LNG facilities, PHMSA anticipates issuing an NPRM to update the LNG regulations by the end of 2023.

As stated during my testimony in response to Rep. Brian Babin of Texas, PHMSA has already initiated planning for the establishment of the Center. Since then, PHMSA has also hosted a public meeting on the subject (in May), made visits to potential locations for the center, and reassigned a senior leader, with considerable LNG experience, to oversee the planning, creation and management of the Center.

Question 20. Do you support efforts to streamline environmental reviews at PHMSA in a way that advances projects without impacting environmental protections, including through applying One Federal Decision, which is bipartisan-supported and applies to other transportation modes?

ANSWER. Congress enacted NEPA (42 U.S.C. 4321 et seq.) and the Pipeline Safety Laws (49 U.S.C. 60101 et seq.) to mitigate environmental and safety impacts. PHMSA is charged by Congress with carrying out these laws in an efficient manner. To do that, we have employed mechanisms like concurrent reviews and close coordination with other federal agencies. When environmental reviews have not been fulsome, the agency's work has actually been slowed down by litigation, which is why the agency has increased its environmental review capacity. The agency is also implementing recent changes to NEPA pursuant to the Fiscal Responsibility Act, see e.g. https://ceq.doe.gov/laws-regulations/fra.html.

Question 21. There are roughly 5,500 hundred miles of CO2 pipeline in this country. These pipelines have safely transported carbon dioxide in a dense or supercritical liquid state for decades mostly without incident and are regulated by PHMSA under 49 CFR Part 195. There is a potential need to build out additional pipeline infrastructure to support carbon capture and sequestration efforts that would potentially move CO2 in less-dense, gaseous phases. This has raised the question of whether current regulations are adequate when it comes to the safe construction, operation, and maintenance of carbon dioxide pipelines. PHMSA has also stated publicly that it will be issuing new regulations on the safety of carbon dioxide pipelines.

Question 21.a. How does PHMSA intend to proceed with this rulemaking? Does it propose to publish an Advanced Notice of Proposed Rulemaking (ANPRM) to gather additional data, information and analysis or does it have enough to proceed with a Notice of Proposed Rulemaking (NPRM)?

ANSWER. PHMSA is developing a notice of proposed rulemaking (NPRM), strengthening its regulations in response to the anticipated significant expansion of carbon dioxide pipelines and to implement lessons learned from the February 22, 2020, rupture of a supercritical-phase carbon dioxide pipeline in Satartia, MS.

PHMSA also announced in 2022, new research topics to better determine impact areas for the safer operations of carbon dioxide pipelines. The results of this may help inform the current rulemaking related to carbon dioxide pipelines but congressional attention and collaboration in strengthening oversight related to these issues is also welcome.

PHMSA hosted a public meeting focused on carbon dioxide pipeline safety on May 31–June 1, 2023, in Des Moines, IA, where PHMSA received information from a diverse array of stakeholders and other members of the public that will inform PHMSA's NPRM.

Question 21.b. When does PHMSA anticipate publishing an ANPRM or NPRM?

ANSWER. PHMSA is working on drafting the NPRM and hopes to publish that NPRM in January 2024.

Question 21.c. Has PHMSA identified particular priority issues the agency believes need to be addressed through a potential future rulemaking?

ANSWER. PHMSA is developing proposed amendments to its regulations in response to the anticipated significant expansion of carbon dioxide pipelines and to implement lessons learned from the February 22, 2020, rupture of a supercritical-phase carbon dioxide pipeline in Satartia, MS. Although PHMSA is still determining what specific issues will be covered in the rule, PHMSA discussed several topics related to carbon dioxide in its public meeting on carbon dioxide pipeline safety on May 31–June 1, 2023, in Des Moines, Iowa. Information gathered at the public meeting will help inform the rulemaking.

Question 21.d. Is PHMSA considering modifications to existing regulations under 49 CFR Part 192 regarding the transportation of gaseous CO2 in pipelines or would it create a new section?

ANSWER. Consistent with 49 U.S.C. 60102(i), PHMSA is considering whether applying the minimum safety standards in 49 CFR part 195 to gaseous carbon dioxide would help improve safety.

Question 22. How does PHMSA intend to work with CO2 pipeline operators, other Federal agencies and public stakeholders to develop and design dispersion models in the case of a pipeline rupture or substantial leak? Will it incorporate dispersion modelling in CO2 pipeline safety regulations?

ANSWER. PHMSA has observed through the published research on the topic that all models have different advantages and disadvantages. As part of its rulemaking process, PHMSA will seek stakeholder input on the best approach to CO2 dispersion modelling. Stakeholder input will include both public input as well as collaboration with other Federal agencies. For example, PHMSA and DOE's Office of Fossil Energy and Carbon Management are discussing these technical issues at agency sponsored research events and have agreed to co-chair an Interagency Carbon Transport Topic Team. This topic team will include several other interested Federal agencies, share information on dispersion modeling among many other technical topics, resolve issues and track projects to their completion.

PHMSA requires operators to have a process for identifying which pipeline segments could affect a high consequence area, including the use of dispersion modeling (see § 195.452(f)(1)). This modeling is required prior to the occurrence of an emergency and can also be used to inform emergency responders in the event of an actual emergency (see § 195.402(e)(8)).

Question 23. Liquid CO2 pipelines are critical platforms for the biofuel industry in the Midwest. Currently, a number of liquid CO2 projects are in development, working with PHMSA, states, local governments, and biofuels producers to develop platforms for the capture and storage of millions of metric tons of CO2. Development of these platforms not only sustains the biofuel industry, critical to farmers and the Midwest economy, but also reduces carbon emissions. Liquid CO2 pipelines are definitionally hazardous liquid pipelines and currently must adhere to safety regulations put forward by PHMSA.

Question 23.a. Is PHMSA currently working to ensure the safety of CO2 pipeline projects that would utilize carbon capture from biofuel plants in the Midwest?

ANSWER. Yes. PHMSA is also reviewing the three proposed projects for compliance with the federal regulations in 49 CFR Part 195, PHMSA is working to issue a Notice of Proposed Rulemaking updating current CO2 regulations, and will continue to oversee the safe design, operations, and maintenance of such lines going forward. PHMSA also recently issued new requirements for automatic and remote shut off valves for CO2 (and other hazardous liquid and gas) pipelines (See 87 Fed. Reg 20940), which PHMSA will be implementing with respect to new CO2 pipelines being built.

Question 23.b. Notwithstanding any potential future regulations, would current PHMSA safety regulations for hazardous liquid pipelines apply to any liquid CO2 pipeline currently built in the United States?

ANSWER. The current regulations apply to any pipeline transporting CO2 in a supercritical state as defined by § 195.2.

Question 24. The Federal Government is spending a substantial amount of funding on hydrogen energy as a residential, modal and industrial fuel. However, building out a network fully dedicated to hydrogen fuel pipelines present a host of challenges.

Question 24.a. Given the fact that blending of natural gas with hydrogen in concentrations up to 20 is occurring on a limited basis, does PHMSA have an opinion on the safety of this approach?

ANSWER. PHMSA is evaluating existing requirements for hydrogen and hydrogen-blended pipelines in 49 CFR Part 192 to identify what changes, if any, are needed in its regulations. In addition, PHMSA is supporting research to determine how transporting hydrogen can affect the integrity of pipelines that today carry natural gas.

Question 24.b. What research is being done on hydrogen blends and pipelines?

ANSWER. In FY 2022, PHMSA awarded approximately $6 million in research investments on hydrogen R&D projects. Specifically, under the

Competitive Academic Agreement Program, PHMSA awarded two projects on pipeline infrastructure and modernization for hydrogen networks to two universities. These projects will research the safe transportation and storage of hydrogen via repurposing existing infrastructure used for natural gas transport and underground storage, improving hydrogen leak detection, and characterizing hydrogen-specific pipeline integrity threats.

Question 24.c. Does PHMSA intend to study the safety impacts of blending at concentrations above 20 percent?

ANSWER. Yes. All PHMSA hydrogen R&D projects, as part of their research scope, involve the safety impacts of blending at concentrations over 20 percent. PHMSA is also closely coordinating with the Department of Energy HyBlend initiative which is also studying the impacts of blending at concentrations over 20 percent. Data resulting from both initiatives will inform national standards and support the safe transportation by pipeline and underground storage of hydrogen.

Question 25. How does PHMSA work with standards-establishing bodies like the American Society of Mechanical Engineers (ASME) and the American National Standards Institute to develop pipeline and operational safety standards? How often does PHMSA review and incorporate new industry safety standards by reference? What is its typical timeline for doing so?

ANSWER. PHMSA employees participate in meetings held by national standards developing organizations (SDO) that address the design, construction, operation, maintenance, inspection, and repair of pipeline facilities. PHMSA's subject matter experts represent the agency, participate in discussions and technical debates, register opinions, and vote in accordance with the procedures of the SDOs at each stage of the standards development process (unless prohibited from doing so by law). PHMSA participates in this process to ensure that the agency's safety priorities are considered and to avoid the need to develop separate, government-unique standards. PHMSA also regularly reviews updated editions of currently referenced consensus standards and amends the regulations to partially or fully incorporate updated standards that will enhance or maintain pipeline and environmental safety. Previous rules that incorporated updated consensus standards by reference were published on May 24, 1996, (61 FR 26121); February 17, 1998, (63 FR 7721); June 14, 2004, (69 FR 32886); June 9, 2006, (71 FR 33402); February 1, 2007, (72 FR 4655 (correction)); August 11, 2010, (75 FR 48593); January 5, 2015, (80 FR 168); and August 6, 2015, (80 FR 46847 (correction)). As of late, PHMSA has issued two NPRMs (January 15, 2021 (86 FR 3938) and

August 29, 2022 (87 FR 52713)) that will update or otherwise address 60 percent of PHMSA's standards that are currently incorporated by reference.

Question 26. PHMSA has undertaken a number of previous rulemakings regarding improvements to state One Call systems.

How would PHMSA rate the effectiveness of these rulemakings in reducing excavation-related damages?

ANSWER. PHMSA's Excavation Enforcement Final Rule issued July 13, 2015, established standards for State damage prevention enforcement programs creating accountability at the State level for effective enforcement of State excavation damage prevention laws. Prior to this rule, the enforcement of one-call laws was strictly a stakeholder-driven, state-by-state approach, and many states were unable to show any consistent or demonstrative enforcement of their laws, even when there were violations of existing State laws. Accordingly, in 2016, when the first round of State evaluations was conducted to determine the adequacy of State damage prevention enforcement programs, 25 of the States failed to meet the requirements of 49 CFR 198.55 and were deemed "inadequate." Currently, the number of States that are inadequate is three.

This rule has helped improve States' enforcement of damage prevention laws. The number of excavation damages per thousand notifications of excavation (One-Call Tickets) has generally trended downward, from 3.0 in 2015 to 2.55 in 2022 indicating improvement in reducing damages when a notification to excavate was made. However, the actual number of excavation damages continues to increase from 81,974 in 2015 to 92,006 for 2022 a 12% increase indicating more work to drive improvement is necessary to address this threat to pipeline and public safety.

Question 27. PHMSA recently accepted applications for the Natural Gas Distribution Infrastructure Safety and Modernization (NGDISM) grant program authorized in the Infrastructure Investment and Job Act of 2021. This program provides $200 million annually to municipal utilities to repair, rehabilitate or replace high-risk distribution pipelines.

Question 27.a. How will PHMSA prioritize applications?

ANSWER. For the FY 2022 funding round, PHMSA considered each of the statutory criteria, and prioritized high-risk, actively leaking legacy natural gas distribution infrastructure with a specific emphasis on benefiting underserved rural and urban communities (among other considerations).

Question 27.b. How will it weigh statutory criteria in terms of what weight will be given to environmental justice criteria?

ANSWER. Please see PHMSA's response to Question 29.

Question 27.c. Can you please list the number of positions that have been filled along with their associated titles and GS-level that have been onboarded to implement this program, as well as any vacant positions?

ANSWER. All 9 positions are currently filled.

- Program Director, GS–15
- Grants Attorney, GS–14
- Senior Grants Management Specialist, GS–14 (Team Lead)
- Grants Management and Acquisitions Specialist, GS–14
- Grants Management Specialist, GS–13 (3 positions)
- Program Specialist, GS–13
- Environmental Protection Specialist, GS–14

Question 27.d. When will PHMSA issue its first awards under this program?

ANSWER. PHMSA announced its first awards on April 5, 2023.

Question 27.e. When will it issue its notice of funding opportunity for the second cycle of funding?

ANSWER. PHMSA issued its combined FY23 and FY 24 Notice of Funding Opportunity on May 23, 2023.

Question 28. The President's FY 2024 budget for PHMSA references "equity" roughly thirty times, including funding related to "equity." Please provide a specific definition of "equity" and explain how it relates to pipeline safety and PHMSA's core mission of pipeline safety.

ANSWER. For clarity, PHMSA also has a safety mandate pertaining to the regulation of transporting hazardous materials. Executive Orders 13985 and 14901, issued by President Biden, define equity as "the consistent and systematic fair, just, and impartial treatment of all individuals, including individuals who belong to underserved communities that have been denied such treatment, including persons who live in rural areas, and persons otherwise adversely affected by persistent poverty or inequality." Equity considerations directly relate to PHMSA's core mission as data analyses comparing pipeline incident rates in socially vulnerable areas indicate a disproportionately higher incident rate in some overburdened and underserved communities. PHMSA will use data to focus inspection efforts on increasing safety in areas with higher incident rates, ensure rulemakings and agency decisions affect the safety of communities regardless of geography or income level, and increase community outreach efforts to raise awareness and education in areas with higher rates of incidents. As an example, aging and higher risk infrastructure are found disproportionately in rural communities and older parts of cities and towns, which often coincide with lower income neighborhoods.

Question 29. The President's FY 2024 budget for PHMSA references "environmental justice" multiple times. Can you please specifically define "environmental justice" and describe how it applies to PHMSA's mission of providing pipeline safety?

ANSWER. Presidential Executive Order 14096 of April 21, 2023, directs Federal agencies to develop an agency-wide environmental justice strategic plan that sets forth the agency's vision, goals, priority actions, and metrics to address and advance environmental justice and to fulfill the directives of this order. EO 14096 defines environmental justice as follows: "Environmental justice" means the just treatment and meaningful involvement of all people, regardless of income, race, color, national origin, Tribal affiliation, or disability, in agency decision-making and other Federal activities that affect human health and the environment so that people: (i) are fully protected from disproportionate and adverse human health and environmental effects (including risks) and hazards, including those related to climate change, the cumulative impacts of environmental and other burdens, and the legacy of racism or other structural or systemic barriers; and (ii) have equitable access to a healthy, sustainable, and resilient environment in which to live, play, work, learn, grow, worship, and engage in cultural and subsistence practices.

PHMSA looks to this order, which builds on and complements Executive Order 12898, and policy objectives of Federal regulation to ensure its regulatory oversight activities advance environmental justice for all, including in evaluating and addressing, as appropriate, disproportionate and adverse benefits and burdens and advancing the goal of meaningful involvement. Following statutory requirements, environmental justice has, therefore, long been an important consideration in PHMSA regulatory oversight of ensuring pipeline safety, addressed explicitly in historical PHMSA pipeline safety rulemakings, special permit reviews, and other activities. Often the procedural vehicle for consideration of environmental justice has been PHMSA's consideration of environmental and public safety impacts pursuant to NEPA.

Question 30. We understand there was great interest in the NGDISM Program, but PHMSA has not announced its awards for FY 2022 as of March 14, 2023. Further, the grant agreements will not be executed until the National Environmental Policy Act (NEPA) Tier 2 Environmental review process is completed in 2023. Smaller utilities often do not have the resources dedicated to devote staff to grant writing and therefore may seek to learn from the first round of grant funding awarded in order to use their resources more efficiently. However, if applications for the second cycle of funding are due before the first cycle funding is awarded, they may miss out on this learning opportunity.

Question 30.a. How is PHMSA working to address potential lost opportunity and working to ensure smaller utilities are able to use the first cycle of awards to use their resources effectively in applying for the program?

ANSWER. The first round of awards was announced on April 5, 2023. PHMSA issued the second NGDISM Notice of Funding Opportunity on Grants.gov on May 23, 2023, with an August 4, 2023 deadline. PHMSA believes that applicants who were not successful in the first round had sufficient opportunity to review their application status letters to determine whether to apply for the second round of funding. PHMSA received 184 applications for its FY 2023 NOFO with funding requests that totaled $1.8 billion. Many applicants from the first round elected to re-apply. To assist smaller utilities with applying for this grant, PHMSA made updated its FY 2023 NOFO to make it more user friendly and provided examples in certain areas to assist applicants. PHMSA has also provided extensive technical assistance via FAQs, webinars, debriefs, and outreach. Going forward, PHMSA will promote DOT's Thriving Communities Program as a resource on its website and inform stakeholders that this program is available for grant application assistance.

Question 30.b. What feedback will be provided after the 2022 grant awards have been provisionally announced so as to allow smaller municipal gas systems to learn how to improve their applications?

ANSWER. PHMSA informed unsuccessful applicants by letter about their application rating and any key factors that impacted their application. Additionally, PHMSA offered applicants the opportunity to discuss questions about their application, including steps they can take to improve their application. In total, 56 of the 161 unique applicants requested and received debriefs. PHMSA concluded these debriefs in June 2023. Finally, PHMSA offered three public webinars to discuss lessons learned and provided examples of best practices and common application pitfalls. PHMSA accepted questions during all three FY 2023 webinars and provided a FY 2023 FAQ document with responses to those questions. PHMSA also continues to accept questions from applicants by email (PHMSAPipelineBILGrant@dot.gov) and phone 202–366–7652. All resource documents are available online at https://www.phmsa.dot.gov/about-phmsa/working-phmsa/grants/pipeline/natural-gas-dis-tribution-infrastructure-safety-and-modernization-grants.

Question 31. In your testimony, you discuss the importance of cybersecurity for pipelines and that PHMSA is working to hire cybersecurity specialists. We understand the Transportation Security Agency (TSA) has the lead on pipeline security and is working on cybersecurity regulations for pipelines.

Question 31.a. Given that TSA has the lead on cybersecurity enforcement and PHMSA faces staffing shortages for pipeline safety personnel, how will hiring new cyber specialists benefit PHMSA's pipeline safety mission?

ANSWER. As stated during my testimony on March 8, 2023, PHMSA's safety oversight of pipeline control rooms forms a nexus with TSA's cybersecurity oversight, the Cybersecurity and Infrastructure Security Agency's (CISA's) role as the national coordinator for critical infrastructure security and as a cybersecurity agency, and DOE's national energy management. The 2021 cyber-attack on Colonial Pipeline demonstrated how critical it is for a whole-of-government approach to safeguard our Nation's critical infrastructure. PHMSA regulates, inspects, and enforces operational technology (OT) used in the communications and supervisory control and data acquisition (SCADA) systems for pipeline, liquefied natural gas and underground natural gas storage facilities. PHMSA's two new cyber specialists will join a small team of PHMSA SCADA and control room experts that routinely inspect the 240–270 control rooms spread across the U.S. and Canada. The specialists will be able to leverage their cyber expertise during the OT portion of CRM inspections, identifying violations and potential cyber-related concerns. In addition, the specialists will provide onsite accident and special investigation assistance. Cyber specialists will look at specific incidents to identify a potential cyber component when reviewing control room details, controller responses, and data associated with a pipeline spill or release into the environment.

Pipeline control rooms are where a cyber-attack could create significant safety hazards to the public and threaten energy reliability, as demonstrated by the Colonial Pipeline hacking incident. PHMSA has oversight of the operator's emergency response to any cyber-attack that impacts the safe operation of the pipeline. Further, PHMSA has safety oversight of a pipeline start-up and return to normal operations, including manual operations should commodity movement be required. The PHMSA inspection team, including those with combined cyber and pipeline operation expertise, are the federal government's regular "in-the-field" presence and will be a critical component of safely restoring pipeline operations should U.S. pipeline infrastructure experience another cyber-attack. It is important to note that PHMSA routinely collaborates with TSA and CISA and will refer any identified cyber risks to TSA for their review and action.

Question 31.b. Can you please list the number of these cybersecurity positions that have been filled along with their associated titles and GS-level? Can you please list the number of vacant positions PHMSA is planning to hire, along with their associated titles and GS-level?

ANSWER. PHMSA will soon advertise two GS–14 General Engineer positions with a program title of "Program Manager." The best qualified candidates for these two positions will have pipeline engineering and control room management expertise combined with a sound cyber skill set. These individuals will be CRM inspectors and provide a unique skill set with understanding of pipeline safety standards, pipeline OT and the potential impact that cyber may have on pipeline OT.

Question 31.c. How will you ensure PHMSA does not duplicate the efforts of the TSA?

ANSWER. PHMSA's safety oversight forms a nexus with TSA's cyber authority and the two agencies have adjacent but different foci. PHMSA regulates, inspects, and enforces operational technology (OT) used in the communications and supervisory control and data acquisition (SCADA) systems for all pipeline, liquefied natural gas (LNG) and underground natural gas storage (UNGS) facilities. PHMSA performs routine, onsite inspections of all pipeline control rooms and connects pipeline operators with TSA when potential cyber risks are identified. PHMSA also routinely investigates both major and minor pipeline, LNG, and UNGS incidents, as well as abnormal events. PHMSA field inspections include onsite observation of OT components such as PLCs (program logic controllers) in-situ, such as in remote, rural settings.

PHMSA evaluates SCADA controller training and human factors such as fatigue. These issues have a cyber nexus as a fatigued, or inadequately trained, controller can make risky OT decisions, potentially missing pressure and flow indications of a leak or abnormal operations.

PHMSA exerts is safety and regulatory authority on all pipeline, LNG and UNGS facilities; Currently, TSA efforts are limited and constrained to certain operators. PHMSA and TSA collaborate under the guidance of Memorandum of Understanding (MOU) that promotes communication, efficiency, and nonduplication of effort.

Question 32. Deputy Administrator Brown, on November 4, 2020, PHMSA published a set of draft frequently asked questions (FAQs) in the Federal Register under docket number PHMSA–2019–0199 attempting to delineate where PHMSA and the Occupational Safety and Health Administration (OSHA) assert jurisdiction to perform inspection and enforcement for midstream processing facilities in which there is overlapping authority by the two respective regulatory bodies. While PHMSA extended the comment period on that proposal to February 2022, to date, PHMSA has not finalized this guidance or taken any other regulatory action to address this issue. I am aware of confusion and other challenges facing the regulated community as a result of the lack of clarity around this issue.

Question 32.a. Will PHMSA take action this year to finalize this guidance or perhaps propose some other remedy to this important issue? If so, when? If not, why?

ANSWER. PHMSA anticipates finalizing the guidance relative to the jurisdiction of midstream processing facilities later this year.

Question 32.b. In the absence of regulatory action, how will PHMSA provide clarity to the regulated community around this issue?

ANSWER. PHMSA believes that the jurisdictional clarity provided by the finalized midstream FAQ's will negate the need for any further regulatory action.

Question 33. The PIPES Act of 2016 convened a Voluntary Information Sharing System working group that recommended Congress authorize such a program to share pipeline safety information while providing confidentiality for disclosures.

Question 33.a. Does PHMSA believe there are benefits to such a program and how might it use such a system to improve pipeline safety across the United States pipeline network?

ANSWER. PHMSA acknowledges potential benefits to the Voluntary Information Sharing System (VIS), if thoughtfully crafted, including the potential for:

- Serving as a trusted repository of high-volume, high-quality data and information that would advance pipeline safety and lead to opportunities for reducing pipeline accidents and incidents to achieve the goal of "zero."
- Enhancing Safety Management Systems.
- Providing technical support for service providers' technology investments to improve technology performance.
- Determining gaps in pipeline information to drive continuous improvement.

Question 33.b. How might Congress or PHMSA address concerns about the sharing of proprietary information by companies who actively participate in a voluntary information sharing program?

ANSWER. Industry may be more incentivized to participate in the program if their concerns about protecting proprietary and confidential information could be better addressed within a VIS program. PHMSA welcomes the opportunity to working with Congress on ways to address these concerns.

Question 34. PHMSA recently started asking pipeline operators during inspections to fill out the "Racial Equity and Support for Underserved Communities—New Racial Equity and Support for Underserved Communities Form 1." PHMSA stated the justification for asking these questions is Executive Order 13985, Advancing Racial Equity and Support for Underserved Communities Through the Federal Government. To aid in the response of questions, PHMSA inspectors pointed to a Climate and Economic Justice Screening Tool map that was put together by the Council on Environmental Quality. We have several questions on this topic:

Question 34.a. What was the process that PHMSA used to draft this form and send it out to pipeline operators? Were other Federal agencies, including the Office of Management and Budget (OMB), consulted? Did PHMSA obtain approval from OMB to conduct this information collection exercise based on Paperwork Reduction Act requirements?

ANSWER. In 2022, PHMSA performed a data analysis comparing incident occurrence locations relative to the Centers for Disease Control (CDC) Social Vulnerability Index (SVI) and found that some SVI areas may be disproportionately impacted by natural gas distribution incidents. Discussions with a variety of stakeholder groups—including state regulators and industry groups—revealed a value in raising awareness and improving transparency about the location of incidents, including for historically underserved communities. PHMSA data shows that a good amount of underserved communities are at significantly greater risk of pipeline failures due to aging infrastructure and other factors. To help with this, PHMSA initially formulated seven questions to ask operators to raise awareness and thoughtful consideration among the industry and regulators regarding EO 13985. PHMSA has since converted the questions into informational points for its inspectors to facilitate discussions with operators during inspections regarding equity or impacts on underserved communities. The points not only help us communicate that we are seeking to advance protection for all, including in these areas, but also helps us understand what efforts operators are taking to address safety risks for local communities. Similar to PHMSA's other inspection resource materials, the original questions were not submitted for OMB approval, as each inspection, including questions, are uniquely designed for each pipeline system.

Question 34.b. Did PHMSA submit this form to the Federal Register and institute a formal comment process? If not, why?

ANSWER. PHMSA did not submit the questions to the Federal Register for comment. PHMSA has converted the questions into informational points in a brochure.

Question 34.c. To which pipeline operators did PHMSA submit this form and how were they selected?

ANSWER. PHMSA had discussed the questions with companies that it interacts with, including during safety inspections. PHMSA has converted the questions into informational points in a brochure.

Question 34.d. The form references "Disadvantaged Community (DAC) datasets." How would you explain what a DAC is and what criteria is used to establish it? Additionally, are DACs used in any part of existing PHMSA regulation?

ANSWER. PHMSA initially applied the acronym DAC to refer to various datasets that include data components that PHMSA deemed relevant to social disadvantage such as components of the Center for Disease Control's (CDC) Social Vulnerability Index (CDC/ATSDR Social Vulnerability Index (SVI) and many others referenced on the Department of Transportation's public facing website at: https://www.transpor tation.gov/ grants/dot-navigator/federal-tools-determine-disadvantaged-community-status. These tools are publicly available and can assist stakeholders in their own data analysis. The Pipeline Social Equity Tool was published on August 8,2023 and does not use the SVI. It uses the DOT Equitable Transportation Community (ETC) Explorer Tool data. DAC PHMSA—Pipeline Social Equity Tool (arcgis.com).

Question 34.e. Has PHMSA incorporated any aspects of Executive Order 13985 into its regulations?

ANSWER. PHMSA has not amended its regulations to incorporate the text of Executive Order 13985. However, PHMSA has addressed, at a high level, the application of Executive Order 13985 within several of its recent rulemakings. See, e.g., 86 FR 63266, 63291 (Nov. 15, 2021). PHMSA also has, consistent with the Federal policy objective announced in Executive Order 13985 and where appropriate, identified equity benefits expected from recent safety-enhancing rulemakings just as it recognizes safety, economic, and environmental benefits from its rulemakings as appropriate.

Question 34.f. Does PHMSA have any plans to change its regulations to address disadvantaged communities? If so, can you please specify which Regulation Identifier Numbers (RIN) you plan on using?

ANSWER. PHMSA is still analyzing the safety data and currently has no new specific RINs on the matter.

Question 34.g. Does PHMSA believe safety and environmental expectations should be different in DAC areas as opposed to non-DAC areas? If so, can you please stipulate how and why?

ANSWER. PHMSA believes safety and environmental expectations and performance should be commensurate in both DAC and non-DAC areas. This is why potentially disparate safety performance is concerning and why PHMSA seeks to raise awareness of the potential issue.

Question 34.h. If PHMSA believes there should be different expectations as stipulated in the above question, can you please specify which regulations govern this process?

ANSWER. As stated above, PHMSA believes expectations should be the same.

Question 34.i. What does PHMSA intend to do with the information gleaned from the answers to the form mentioned above?

ANSWER. PHMSA's intention is to raise awareness through discussion, and any information gathered will be used only to document staff effort to reach stakeholders.

Question 34.j. The Executive order defines "equity" as the "consistent and systematic fair, just and impartial treatment of all individuals." Do you believe that the current set of pipeline regulations included in 49 CFR Parts 190–199 are adequate regarding equity as defined above? If not, why?

ANSWER. While the intent of PHMSA regulations is designed to accomplish equity as defined in the EO, PHMSA believes empirical data is more valuable than interpreting regulatory intent.

Question 34.k. The Department of Transportation recently sent to the Federal Register a notice regarding its updated Transportation Disadvantaged Census Tracts Tool asking for input from stakeholders. How does PHMSA plan on using this tool?

ANSWER. On August 8, 2023, PHMSA published an interactive map called the Pipeline Social Equity Tool. This public tool allows stakeholders to evaluate safety performance compared to the Department of Transportation's

Equitable Transportation Community (ETC) Explorer Tool. PHMSA has not yet identified other specific uses for the Tool.

Questions from Hon. Donald M. Payne, Jr. to Tristan Brown, Deputy Administrator, Pipeline and Hazardous Materials Safety Administration (PHMSA)

Question 1. In 2022, there were more than 280 significant or serious incidents and this year we have already had two fatalities as a result of pipeline incidents. Meanwhile, your testimony states that certain operators are negotiating for reduced civil penalties and other operators are delaying paying the civil penalties they owe. Which pipeline operators are negotiating for lesser civil penalty payments and which ones have not paid what they owe? For the operators who have not paid, how long have you been awaiting their payment?

ANSWER. Sec. 108 of the 2020 PIPES Act directs PHMSA to allow the respondent in an enforcement proceeding to request the use of a consent agreement. It further directs PHMSA to allow the respondent and the agency to convene at least one meeting to explore the possibility of settlement or simplification of the issues. On average over the last two years, PHMSA issued 47 proposed penalties and consistent with Sec. 108, every operator was eligible to request the use of a consent agreement and to meet with PHMSA for settlement discussions. Since the beginning of 2018, PHMSA, as part of settlement negotiations, issued penalties lower than proposed in 24 pipeline enforcement cases, as noted below:

CPF NUM	*OPERATOR*	*PROPOSED PENALTIES*	*ASSESSED PENALTIES*	*OPENED DATE*	*CONSENT ORDER DATE*
42022017NOPV	DENBURY GULF COAST PIPELINES, LLC (32545).	$3,866,734	$2,868,100	5/26/22	3/24/23
12022038NOPV	KINDER MORGAN LIQUID TERMINALS, LLC (26041).	$455,200	$165,700	7/1/22	2/3/23
32022019NOPV	TEXAS GAS TRANSMISSION, LLC (19270).	$474,300	$237,800	1/21/22	7/1/22
32022018NOPV	TALLGRASS PONY EXPRESS PIPELINE, LLC (39043).	$55,200	$44,800	1/19/22	5/16/22

CPF NUM	*OPERATOR*	*PROPOSED PENALTIES*	*ASSESSED PENALTIES*	*OPENED DATE*	*CONSENT ORDER DATE*
32021046NOPV	TALLGRASS PONY EXPRESS PIPELINE, LLC (39043).	$437,300	$385,500	11/23/21	5/10/22
32021043NOPV	TALLGRASS INTERSTATE GAS TRANSMISSION, LLC (1007).	$359,900	$325,382	12/7/21	4/13/22
32021045NOPV	TALLGRASS POWDER RIVER GATEWAY, LLC (39963).	$58,400	$25,920	11/5/21	3/29/22
32021049NOPV	DAPL–ETCO OPERATIONS MAN-AGEMENT, LLC (39205).	$93,200	$20,000	7/22/21	1/11/22
320215008 .	HESS ND (39065)	$127,600	$82,000	3/1/21	12/1/21
32021022NOPV	EXXONMOBIL PIPELINE CO (4906)	$58,200	$38,900	5/18/21	10/28/21
52021022NOPV	SFPP, LP (18092)	$2,231,779	$1,493,200	5/12/21	10/22/21
32021027NOPV	NORTHERN NATURAL GAS CO (13750).	$56,900	$46,700	4/28/21	9/10/21
320205024 .	SPIRE MISSOURI INC. EAST (11032).	$139,800	$62,600	11/30/20	4/9/21
42020018NOPV	TEXAS EASTERN TRANSMISSION, LP (SPECTRA ENERGY PARTNERS, LP) (19235).	$36,200	$18,100	12/9/20	4/8/21
320205004 .	PEMBINA COCHIN LLC (32258)	$187,200	$172,800	3/19/20	3/18/21
320205020 .	SINCLAIR TRANSPORTATION COMPANY (15156).	$97,100	$67,700	10/13/20	2/5/21
520192001 .	HILCORP ALASKA, LLC (32645)	$198,700	$0	1/24/19	5/18/20
320196006	CRESTWOOD MIDSTREAM PART-NERS LP (39368).	$236,100	$200,000	11/22/19	5/13/20
120190004 .	RICHMOND, CITY OF (17360)	$80,500	$0	4/22/19	4/22/20
420145025 .	CENTURION PIPELINE L.P. (31888)	$165,600	$92,000	9/30/14	1/31/19
120160005 .	RICHMOND, CITY OF (17360)	$51,800	$0	10/13/16	11/30/18
320135014 .	ONEOK NGL PIPELINE, LLC (32109).	$559,100	$550,400	5/13/13	10/23/18
320135015 ..	ONEOK NGL PIPELINE, LLC (32109).	$230,800	$0	5/13/13	10/23/18

CPF NUM	*OPERATOR*	*PROPOSED PENALTIES*	*ASSESSED PENALTIES*	*OPENED DATE*	*CONSENT ORDER DATE*
320135020	ONEOK NGL PIPELINE, LLC (32109).	$45,700	$22,500	7/3/13	10/23/18

The following two enforcement cases are in an extended non-payment status:

- Idaho Pipeline Corp has not paid $49,000 penalty that PHMSA assessed in an August 9, 2019, Final Order (CPF 5–2018–6015).
- Bohrenworks LLC has not paid $209,002 penalty that PHMSA assessed in a December 2, 2019, Final Order (CPF 5–2019–0018E). Bohrenworks is an excavator.

Question 2. The 2020 PIPES Act allowed PHMSA a number of recruitment and retention incentives, including special pay rates permitted under section 5305 of title 5, United States Code. How has having this authority assisted PHMSA to recruit and retain inspection and enforcement personnel? Are there additional barriers to ensuring PHMSA has the resources it needs?

ANSWER. In recognition of the critical nature of these positions, the PIPES Act of 2020 authorized PHMSA to provide recruitment and retention incentives such as tuition assistance, student loan repayment, and special pay rates to improve efforts to attract and retain pipeline engineers and inspectors. In 2021, PHMSA engaged an outside contractor to determine the parameters for increasing its special pay rates—including for example the geographic areas and pay grades. In FY22, PHMSA received additional funding to provide financial incentives such as increasing the pay rates. In August of 2022, PHMSA/DOT submitted to the Office of Personnel Management (OPM) for approval an increase in the special pay rates for engineer/inspectors. PHMSA received OPM approval to increase special pay rates for some engineer inspectors on April 17, 2023, and promptly applied this to increase to existing employees and openings. While awaiting approval, PHMSA increased the number of open/available positions for which it is hiring. In 2023, PHMSA also proposed establishing additional incentives to retain pipeline engineers/inspectors. Upon approval of these new incentives, the agency hopes it will experience a significant boost in its rate of onboarding and retention of inspectors.

Question 3. California state law (2022 SB–905) prohibits construction of any new pipelines for carbon dioxide, including for new carbon capture and storage projects, until PHMSA promulgates a final rulemaking regarding minimum federal safety standards for transportation of carbon dioxide by pipeline. Can you confirm that PHMSA is working to finalize this rulemaking, and provide a publication date for the Notice of Proposed Rulemaking?

ANSWER. PHMSA is developing a proposed rule updating its regulations in response to the anticipated significant expansion of carbon dioxide pipelines and to implement lessons learned from the February 22, 2020, rupture of a supercritical-phase carbon dioxide pipeline in Satartia, MS. PHMSA anticipates publishing an NPRM by January 2024.

Questions from Hon. Mike Bost to Tristan Brown, Deputy Administrator, Pipeline and Hazardous Materials Safety Administration (PHMSA)

Question 1. Does PHMSA oppose acts to damage or destroy an interstate natural gas or hazardous liquids pipeline facility?

ANSWER. Yes, as noted in my response to Congressman Stauber, and PHMSA personnel work closely with the Transportation Security Administration (TSA) to advance pipeline security, which is under the jurisdiction of the TSA.

Question 2. Does PHMSA consider acts to damage or destroy an interstate natural gas or hazardous liquids pipeline facility a legal expression of 1st Amendment free speech rights?

ANSWER. PHMSA does not condone any act to damage or destroy a pipeline facility. Knowingly and willfully damaging or destroying a pipeline facility is a criminal violation of 49 U.S.C. 60123.

Question 3. Does PHMSA consider acts to damage or destroy an interstate natural gas or hazardous liquids pipeline facility an appropriate form of protest against climate change?

ANSWER. PHMSA does not condone any act to damage or destroy a pipeline facility. Knowingly and willfully damaging or destroying a pipeline facility is a criminal violation of 49 U.S.C. 60123.

Question 4. Does PHMSA oppose publications or other forms of media, such as movies, which encourage the damage or destruction of an interstate natural gas or hazardous liquids pipeline facility?

ANSWER. PHMSA does not condone any act to damage or destroy a pipeline facility. Knowingly and willfully damaging or destroying a pipeline facility is a criminal violation of 49 U.S.C. 60123.

Question 5. Does PHMSA consider dangerous certain acts to tamper with or disrupt the operation of an interstate natural gas or hazardous liquids pipeline facility, such as valve-turning?

ANSWER. Only trained personnel with explicit direction and authority of the pipeline operator should ever perform activities on a pipeline facility.

Question 6. Does PHMSA agree with the statement expressed by former Pipeline Safety Trust (PST) Executive Director Carl Weimer that, "[c]losing valves on major pipelines can have unexpected consequences endangering people and the environment?"

ANSWER. Only trained personnel with explicit direction and authority of the pipeline operator should ever perform activities on a pipeline facility.

Question 7. Does PHMSA agree with former PST Executive Director Weimer that actions by activists to close valves on major pipelines were a "dangerous stunt?"

ANSWER. Only trained personnel with explicit direction and authority of the pipeline operator should ever perform activities on a pipeline facility.

Question 8. Does PHMSA support a Federal prohibition on activities, such as closing valves on major pipelines, that are dangerous but do not result in damage or destruction?

ANSWER. Only trained personnel with explicit direction and authority of the pipeline operator should ever perform activities on a pipeline facility.

Questions from Hon. Donald M. Payne, Jr. to Andrew J. Black, President and Chief Executive Officer, Liquid Energy Pipeline Association (LEPA)

Question 1. How much in annual revenue did LEPA members earn in 2022?

ANSWER. LEPA does not collect or hold information on the revenue of its member companies. LEPA focuses on policies and programs that improve pipeline safety and the public's perception of the benefits of pipelines. LEPA also advocates for the liquids pipeline industry at FERC, where the focus is on just and reasonable tariff rates and terms of service. Public sources of information on pipeline operator revenue would provide some information, although this data would be incomplete. Several major members of LEPA are privately owned and do not report their financial information publicly. Several other major members are subsidiaries of larger organizations with multiple business units and generally do not publicly break out liquids pipeline revenues separately.

Question 2. How much of your members' product was exported to other countries? ANSWER. For the most part, LEPA member companies do not own the product they transport. Similar to FedEx or UPS, pipeline companies deliver liquid energy products owned by other parties from one destination to another on behalf of shippers. The ultimate destination of products delivered by LEPA members, to the extent they are beyond the U.S. pipeline network, are usually unknown to LEPA and its members. This model is similar to the rail and trucking industries, where trains and trucks deliver packages and goods owned by other parties from contracted pick-up to destination points. The ultimate end point of those shipments is not necessarily known to the rail and trucking companies.

Question 3. Could you please explain what statutory prohibitions there are on the industry creating its own voluntary information sharing system?

ANSWER. There are no statutes which directly prohibit the sharing of pipeline safety information. However, legal considerations and other risks deter or limit free disclosure of information by infrastructure owners to a voluntary information sharing system. For example, companies will not share safety information which has a security component if it would pose a threat to public safety if publicly released. Companies will also not disclose to the public information which might be used against them in pending or threatened litigation. This is separate and apart from information subject to inspection and enforcement action by a regulator. Congress recognized that legal liabilities discourage safety sharing when it authorized a VIS-like program for air carriers administered by the FAA. Indeed, the success of the FAA safety sharing program resulting from the Congressional and regulator safe harbor granted to air carriers was the inspiration for Congress suggesting application of the program to the pipeline sector. If Congress wants pipeline safety to benefit from voluntary information sharing, it must create the legal safe harbors necessary for such sharing.

Questions from Hon. Donald M. Payne, Jr. to Kenneth W. Grubb, Chief Operating Officer, Natural Gas Pipelines, Kinder Morgan, Inc., on Behalf of the Interstate Natural Gas Association of America (INGAA)

Question 1. How much in annual revenue did INGAA members earn in 2022?

ANSWER. Many of INGAA's member companies have multiple revenue streams in addition to interstate natural gas transmission pipelines. To find the annual revenue stream for each Federal Energy Regulatory Commission ("FERC")-regulated interstate natural gas transmission pipeline, please see each pipeline's FERC Form No. 2 and 2–A, which are publicly available here [https://www.ferc.gov/industries-data/natural-gas/industry-forms/form-2-2a-3-q-gas-historical-vfp-data]. To find the total annual revenue for each INGAA member company for all of its business segments, here is a link to each INGAA member company or their owner's public disclosures and/or press releases:

- BHE GT&S: https://www.brkenergy.com/assets/upload/financial-filing/2022 1231lBHE%20Form%2010K.pdf
- Boardwalk Pipelines: https://loews.com/FileStore/2022-Q4---Earnings-Remarks.pdf
- Cheniere Energy, Inc.: https://www.sec.gov/ix?doc=/Archives/edgar/data/3570/000000357023000042/lng-20221231.htm#ic113 1b9e5c5b41368c2dbfdabf3591 5el19
- DT Midstream: https://s28.q4cdn.com/581450200/files/doclpresentation/2023/01/ DTM-Q4-2022-Earnings-Call-vF.pdf
- DTE Energy: https://ir.dteenergy.com/news/press-release-details/2023/DTE-En-ergy-closes-year-with-strong-financial-results-well-positioned-for-2023/de- fault.aspx
- Eastern Shore Natural Gas: https://investor.chpk.com/static-files/aff97881-7a86- 4f51-a5ab-0768337b06d5
- Enbridge Energy: https://www.enbridge.com/-/media/Enb/ Documents/Investor-Relations/2022/2022lQ4lFinancialsl MDA.pdf?rev=c5741883134 54a118924d6eb033edfcb&hash= 8B9FDDC80A7994E26138F731C077F4EC
- Equitrans Midstream: https://s22.q4cdn.com/743133753/files/doclnews/Equitrans-Midstream-Announces-Full-Year-and-Fourth-Quarter-2022-Results-2023.pdf

- Iroquois Pipeline Operating Company—reported through TC Energy and BHE GT&S
- Kinder Morgan, Inc.: https://s24.q4cdn.com/126708163/files/doclfinancials/2022/q4/KMI-2022-10K-Final-as-Filed-wo-Exhibits.pdf
- Millennium Pipeline Company, LLC—reported through TC Energy and DT Mid- stream
- Mountain West Pipeline—acquired by Williams in Feb. 2023
- National Grid: https://www.nationalgrid.com/document/148586/download
- National Fuel Gas Supply Corporation: https://www. nationalfuel.com/wp-con-tent/uploads/documents/NFG-9.30.2022-Earnings-Release-11032022-PDF.pdf
- NextEra Energy: https://www.investor.nexteraenergy.com/□/media/Files/N/NEE-IR/reports-and-fillings/quarterly-earnings/2022/Q4/2023-0125lNEEQ42022News%20Release%20Final.pdf
- ONEOK, Inc: https://otp.tools.investis.com/clients/us/oneoklinc2/SEC/sec-show.aspx?FilingId=16441050&Cik=0001039684& Type=PDF&hasPdf=1
- Pacific Gas & Electric: https://s1.q4cdn.com/880135780/files/docl financials/2022/q4/PGE-02.23.23-Press-release.pdf
- Sempra LNG: https://www.sempra.com/sempra-reports-fourth-quarter-2022- business-results
- Southern Company Gas: https://s27.q4cdn.com/273397814/files/doclfinancials/ 2022/q4/Earnings-Release-Q4-2022-IR.pdf
- Southern Star Central Gas Pipeline, Inc.—privately owned company that does not report publicly in U.S.
- Spire, Inc.: https://ir.spire.com/sec-filings/all-sec-filings/content/0000950170-23- 008177/0000950170-23-008177.pdf
- TC Energy: https://www.tcenergy.com/siteassets/pdfs/investors/reports-and-fil-ings/annual-and-quarterly-reports/2022/tc-2022-annual-report.pdf
- Tellurian, Inc.: https://ir.tellurianinc.com/financials-filings-and-presentations/an-nual-reports/content/0000061398-23-000011/000 0061398-23-000011.pdf
- UGI Energy Services, LLC: https://www.ugicorp.com/news-releases/news-re- lease-details/ugi-reports-fiscal-2022-results

- WBI Energy Transmission, Inc.: https://d18rn0p25nwr6d.Cloud front.net/CIK-0000067716/c0870b54-f2c5-4d64-88aa-6000c3abd644.pdf
- The Williams Companies: https://d18rn0p25nwr6d.cloudfront.net/CIK-0000107263/828947a0-eef8-45b8-93a2-c623e08108d5.pdf

Question 2. How much of INGAA members' product was exported to other countries?

ANSWER. As you may know, INGAA member companies do not own the natural gas that they transport. Interstate natural gas transmission companies merely transport the natural gas owned and controlled by the pipeline customer (shipper). In an integrated pipeline network (both inter and intrastate), pipeline customers can transport their natural gas over multiple pipelines from various production fields. The interstate natural gas transmission pipeline typically does not know the final end use of the gas that it transports.

INGAA does have member companies which own and operate LNG import and export terminals, including BHE GT&S (Cove Point, MD), Cheniere (Sabine, LA and under construction in Corpus Christi, TX), Kinder Morgan (Elba Island, GA), Sempra (Hackberry, LA), and Tellurian (under construction in Calcasieu Parish, LA). Please see a list of FERC-certificated LNG terminals (both INGAA and non-INGAA members) here [https://cms.ferc.gov/media/north-american-lng-export-terminals-existing-approved-not-yet-built-and-proposed-8]. INGAA does not track the volume of members' exports. Please refer to their websites for further information.

We note that the U.S. Energy Information Administration ("EIA") reports [https://www.eia.gov/dnav/ng/hist/n9133us2m.htm] monthly LNG export volumes and reports [https://www.eia.gov/dnav/ng/hist/n9050us2a. htm] yearly marketed production volumes.

Question 3. What is the starting and average salary, benefits and bonus, and any additional employment incentives for a Kinder Morgan engineer?

ANSWER. Kinder Morgan utilizes a market-based approach whereby pay ranges are established around a market reference point (MRP) to compensate employees within its organization. For pipeline engineering positions, there are 6 job titles that make up the job family currently used by the company with different experience levels and job functions. The average starting compensation for a pipeline engineer at Kinder Morgan can range from \$121,521 to \$172,129, depending on the employee's level of experience and qualifications. This dollar figure includes salary, bonus, and benefits amounting to roughly one third of the total compensation (i.e., health care, payroll taxes, retirement plans, educational reimbursement, and service awards). The average total compensation across the range of all pipeline engineering positions at Kinder Morgan is $208,444 (including the same categories listed earlier).

Question 4. Kinder Morgan has received 39 special permits from PHMSA in the last 10 years and experienced more than 500 accidents or incidents. Have any of those accidents or incidents occurred on pipelines that received a special permit to waive PHMSA's safety regulations?

ANSWER. Kinder Morgan currently has 17 active natural gas transmission class location special permits. No incidents have ever occurred in a class location special permit segment. In the last 10 years, companies operated by Kinder Morgan have experienced 188 natural gas incidents (the remainder of the incidents cited in your question were on our liquids systems). Almost exactly two-thirds of those natural gas incidents, while reportable, were minor and contained within company facilities (e.g., within a compressor or meter station). Additionally, many of Kinder Morgan's pipelines are hundreds of miles long. When there were incidents on pipelines where class location special permit segments exist, there was no impact to the class location special permit segment.

Questions from Hon. Troy E. Nehls to Bill Caram, Executive Director, Pipeline Safety Trust

Question 1. Information sharing is a critical part of the pipeline industry's efforts to advance safety. Since 2016, industry has supported the idea of establishing a formal voluntary information sharing framework, similar to what is utilized by the airline industry. The PIPES Act of 2016 convened a Voluntary Information Sharing System working group that recommended Congress authorize such a program to share pipeline safety information while providing confidentiality for disclosures, but such a provision has been excluded from two previous reauthorization bills.

As the committee considers future safety needs, could you please share your thoughts on whether such a framework would be valuable and should be included in the next pipeline safety reauthorization bill?

ANSWER. The Pipeline Safety Trust believes there is nothing that prevents the pipeline industry from creating an information sharing system if desired, so we question the need for Congressional authorization of such a program. However, we would not oppose a provision provided protections from disclosure, FOIA requests, and other legal processes are reasonable and do not restrict information from the public that is currently available; funding does not pull from existing PHMSA resources; and the governing board is fairly balanced among stakeholders.

Questions from Hon. Donald M. Payne, Jr. To Bill Caram, Executive Director, Pipeline Safety Trust

Question 1. Would you recommend that the committee extend the Technology Pilot Program as mandated by Sec. 104 of the PIPES Act of 2020? If so, do you have specific safety concerns or concerns with the application process? What recommendations do you have to ensure applicants provide a sufficient level of safety during their pilot program?

ANSWER. Given that the Technology Pilot Program would allow companies to operate outside of existing Federal minimum safety standards, we believe the program requires robust safeguards, including an extensive application and approval process, to operate pipelines in our nation's communities without meeting those standards. Regulations around maximum pressure, repair criteria, risk assessments, and many other standards are important to the safety of people and the environment and PHMSA needs to rigorously assess the consequences of any potential regulatory leniency.

The Pipeline Safety Trust would need to assess any proposed changes to the existing program that might accompany a proposed extension to the Technology Pilot Program before deciding to support its adoption.

Question 2. Does the Pipeline Safety Trust support H.R.4261, the "Pipeline Seismic Safety Study Act" raised in the last Congress?

ANSWER. Yes, we support the "Pipeline Seismic Safety Study Act."

Questions from Hon. Mike Bost to Bill Caram, Executive Director, Pipeline Safety Trust

Question 1. Does the PST oppose acts to damage or destroy an interstate natural gas or hazardous liquids pipeline facility?

ANSWER. Yes, The Pipeline Safety Trust opposes activities that create unsafe conditions for people and the environment on pipeline facilities.

Question 2. Does PST consider acts to damage or destroy an interstate natural gas or hazardous liquids pipeline facility a protected expression of 1st Amendment free speech rights?

ANSWER. Acts to damage or destroy a pipeline can create unsafe conditions, and The Pipeline Safety Trust opposes activities that create unsafe conditions for people and the environment on pipeline facilities.

Question 3. Does PST consider acts to damage or destroy an interstate natural gas or hazardous liquids pipeline facility an appropriate form of protest against climate change?

ANSWER. Acts to damage or destroy a pipeline can create unsafe conditions, and The Pipeline Safety Trust opposes activities that create unsafe conditions for people and the environment on pipeline facilities.

Question 4. Does PST oppose publications or other forms of media, such as movies, which encourage the damage or destruction of an interstate natural gas or hazardous liquids pipeline facility?

ANSWER. Unfortunately, this hypothetical question goes beyond our mission and our tiny staff's capacity, but I will reiterate that we oppose activities that create unsafe conditions for people and the environment on pipeline facilities. However, I am unable to comment on the potential constitutional implications of your question.

Question 5. Does PST consider dangerous certain acts to tamper with or disrupt the operation of an interstate natural gas or hazardous liquids pipeline facility, such as valve-turning?

ANSWER. Acts such as valve-turning by untrained individuals can create unsafe conditions on pipeline facilities and the Pipeline Safety Trust opposes activities that create unsafe conditions for people and the environment on pipeline facilities.

Question 6. Do you agree with the statement expressed by former PST Executive Director Carl Weimer that, "[c]losing valves on major pipelines can have unexpected consequences endangering people and the environment?"

ANSWER. Yes, I agree with that statement.

Question 7. Do you agree with former PST Executive Director Weimer that actions by activists to close valves on major pipelines were a "dangerous stunt?"

ANSWER. I am not familiar with the particular circumstances that would contextualize Mr. Weimer's comment but given the level of respect I have for his integrity and judgment, I think it's safe to say I would agree with his assessment at the time.

Question 8. Does PST support a Federal prohibition on activities, such as closing valves on major pipelines, that are dangerous but do not result in damage or destruction?

ANSWER. I believe such activities are already prohibited.

Question 9. Does PST support actions to blow up a natural gas or hazardous liquids pipeline encouraged by the upcoming movie "How to Blow Up a Pipeline" set for release April 7 in theaters across America?

ANSWER. We have not seen the movie nor read the book and do not know if either encourages blowing up pipelines. I will reiterate that the Pipeline Safety Trust opposes activities that create unsafe conditions for people and the environment on pipeline facilities.

Question 10. Does PST consider actions to blow up a natural gas or hazardous liquids pipeline as encouraged by the upcoming movie "How to Blow Up a Pipeline" a danger to people or the environment?

ANSWER. We have not seen the movie nor read the book and do not know if either encourages blowing up pipelines. I will reiterate that the Pipeline Safety Trust opposes activities that create unsafe conditions for people and the environment on pipeline facilities.

Question 11. Does PST condemn the upcoming movie "How to Blow Up a Pipeline?"

ANSWER. We have not seen the movie and have enough real pipeline issues to keep our tiny staff more than busy without spending time dealing with fictional drama meant to entertain. It is always our hope that pipeline companies, law-makers, and regulators, who have the ability to change things safely and rapidly, will deal with the existential threat of climate change in a proactive manner.

Chapter 2

Pipeline Safety: Fully Implementing Key Practices Could Help Improve Transparency of Enforcement Website*

United States Government Accountability Office

Abbreviations

DOT	Department of Transportation
LNG	liquefied natural gas
PHMSA	Pipeline and Hazardous Materials SafetyAdministration
PIPES Act of 2020	Protecting our Infrastructure of Pipelines and Enhancing Safety Act of 2020

Why GAO Did This Study

Accidents on interstate natural gas and hazardous liquid pipelines can release large amounts of these products, potentially injuring or killing people and damaging the environment. In 2007, PHMSA launched the Enforcement Transparency website to address concerns about the transparency of its pipeline safety enforcement actions.

The Protecting our Infrastructure of Pipelines and Enhancing Safety Act of 2020 includes a provision for GAO to review information on pipeline

* This is an edited, reformatted and augmented version of the United States Government Accountability Office Report to Congressional Committees, Publication No. GAO-23-105537, dated November 2022.

In: Pipeline Safety: Background, Practices and Regulatory Changes
Editor: Michael E. Massaro
ISBN: 979-8-89530-058-9

enforcement actions that PHMSA makes publicly available on the internet. This report addresses (1) PHMSA's process to record pipeline safety enforcement information, and what information PHMSA makes publicly available, and (2) the extent to which its pipeline enforcement website aligns with key practices for transparently reporting data.

GAO reviewed relevant statutes, regulations, and agency guidance on recording and publishing pipeline safety enforcement information. GAO compared PHMSA's Enforcement Transparency website with leading practices for transparently reporting government data that GAO previously identified, and interviewed PHMSA officials and stakeholders.

What GAO Recommends

GAO makes three recommendations to improve the transparency and usability of PHMSA's Enforcement Transparency website, specifically to provide complete data in more useful formats, to improve matching with other PHMSA datasets, and to optimize the website's search function. DOT concurred with GAO's recommendations.

What GAO Found

Pipeline and Hazardous Materials Safety Administration (PHMSA) officials inspect pipelines and investigate accidents. These officials can issue a range of enforcement actions if they identify unsafe conditions or non-compliance with laws or regulations. PHMSA records enforcement information in an electronic records system, and uploads documents and related data to its public Enforcement Transparency website. GAO found that this website fully met two of five key practices that GAO identified for transparently reporting open government data and partially met another three. (See figure.) Open data can foster accountability and public trust by providing citizens with information on government activities and their outcomes. PHMSA officials stated that they would like to improve the website for users, but are first focusing on improving the electronic records system. Nevertheless, fully implementing these key practices could assist PHMSA with transparently presenting enforcement information and help users find the information they need.

Key Practices for Transparently Reporting Open Government Data and the Pipeline and Hazardous Materials Safety Administration's (PHMSA) Enforcement Transparency Website

Key practices	Website characteristics
Engage with users	Met. PHMSA officials discuss the use of the website with users, includingpipeline operators and the public, and provide a link on the website for users to submit feedback or questions.
Fully describethe data	Met. PHMSA's website provides labels and descriptions of the enforcementdata and tables. Metadata and licensing information for these data are available on a central Department of Transportation website.
Provide free andunrestricted data	Partially met. PHMSA's website and the data it contains are available to the public at no cost. However, PHMSA does not publish complete data onits website. For example, PHMSA's enforcement documents include variables that indicate whether the pipeline involved is interstate or intrastate, but PHMSA does not make these variables available in the enforcement dataset.
Provide data in useful formats	Partially met. PHMSA allows users to view and download disaggregated data on all enforcement activities. However, users are unable to downloada bulk dataset of all enforcement activities from 2002 to 2022. Additionally, users cannot easily match records from the enforcement website with records in PHMSA's accident dataset.
Facilitate data discovery for users	Partially met. PHMSA facilitates data discovery by providing users with overviews and definitions of the enforcement information published on thewebsite, and provides national summaries of the data. However, PHMSA does not provide an optimized search tool that allows users to filter by discreet variables, such as the regulation violated.

Source: GAO analysis of PHMSA website and key practices. | GAO-23-105537.

November 14, 2022

The Honorable Maria Cantwell
Chair

The Honorable Roger F. Wicker
Ranking Member
Committee on Commerce, Science, and Transportation
United States Senate

The Honorable Frank Pallone, Jr.
Chair

The Honorable Cathy McMorris Rodgers
Republican Leader
Committee on Energy and Commerce
House of Representatives

The Honorable Peter A. DeFazio
Chair

The Honorable Sam Graves
Ranking Member
Committee on Transportation and Infrastructure
House of Representatives

About 360,000 miles of interstate pipelines transport natural gas, oil, and other products across the United States. Although pipelines are relatively safe when compared to transportation alternatives, such as truck and rail, pipeline accidents can release large amounts of these products that can injure or kill people and damage the environment. Within the U.S. Department of Transportation, the Pipeline and Hazardous Materials Safety Administration (PHMSA) oversees safety for pipelines carrying gas and hazardous liquids.[1] PHMSA's oversight includes setting and enforcing the federal minimum pipeline safety standards for the design, construction, operation, and maintenance of these pipelines.

PHMSA officials inspect pipelines, investigate accidents, and can issue a range of enforcement actions, such as Warning Letters and Notices of Probable Violations, if the officials identify unsafe conditions or non-compliance with pipeline safety laws or regulations. PHMSA uploads information on these actions to its Enforcement Transparency website—

[1] PHMSA's general authority is under the Pipeline Safety Laws codified at 49 U.S.C.§ 60101 et seq., and its regulations are located in 49 C.F.R. Parts 190-199.

including documents related to enforcement cases and data on the number of cases opened and closed each year. Pipeline safety groups, industry groups, pipeline operators, and members of the public may use this information to monitor PHMSA's enforcement efforts or to identify trends in pipeline safety issues.

The Protecting our Infrastructure of Pipelines and Enhancing Safety Act of 2020 (PIPES Act of 2020) includes a provision for GAO to review information on pipeline enforcement actions that PHMSA makes publicly available on the internet.[2] This report addresses:

- PHMSA's process to record pipeline safety enforcement information and the information PHMSA makes publicly available, and
- the extent to which PHMSA's pipeline enforcement website aligns with key practices for transparently reporting data.

To describe PHMSA's process to record pipeline safety enforcement information, we reviewed relevant statutes, regulations, and agency guidance. We also reviewed PHMSA's user guides and other documents that describe its electronic records management system, called the Work Management System. Additionally, we interviewed PHMSA officials about their processes to collect and record enforcement information. To describe the enforcement information PHMSA makes publicly available, we reviewed PHMSA's guidance that details which documents it posts on its website and the steps it takes to ensure those documents are posted correctly. We compared this guidance with the statutory requirements for which documents it must make public. We also reviewed PHMSA's website to identify examples of types of documents it makes public. We did not verify that PHMSA has posted every document it is legally required to post for each enforcement case on its website. Rather, we reviewed the website to identify at least one example of all of the document types PHMSA posts according to its guidance.

To determine the extent to which PHMSA's Enforcement Transparency website follows key practices for transparently reporting data, we reviewed this website and the data available on the website, and interviewed agency officials. We compared PHMSA's website to our previous work that identified key practices and actions for transparently reporting data on government websites. Key practices include, for example, providing data in useful formats

[2] Pub. L. No. 116-260, div. R, § 108(a), 134 Stat. 1182, 2221 (2020) (codified at 49 U.S.C.§ 60117(b)).

and engaging with users.[3] Specifically, an analyst reviewed the website, data, and interviews and assessed whether PHMSA had implemented the key actions associated with each key practice. We documented this analysis, and another analyst reviewed and confirmed the results. Based on whether PHMSA's website addressed key actions, we scored each key action as Fully Met, Partially Met, or Not Met. We also interviewed pipeline stakeholders about their use of PHMSA's pipeline enforcement data and the website, and any related challenges.[4] We selected stakeholders to interview who captured a range of perspectives, including those from industry, safety and environmental organizations, and state governments. Stakeholder views cannot be generalized to represent the views of all stakeholders, but they provide valuable perspectives on PHMSA's Enforcement Transparency website.

Additionally, we analyzed PHMSA data to describe the number and type of enforcement cases that PHMSA initiated from 2010 to 2020. See appendix 1 for information on the steps we took to determine that the data were reliable for our purposes and on the methodology and results of this analysis.

We conducted this performance audit from November 2021 to November 2022 in accordance with generally accepted government auditing standards. Those standards require that we plan and perform the audit to obtain sufficient, appropriate evidence to provide a reasonable basis for our findings and conclusions based on our audit objectives. We believe that the evidence obtained provides a reasonable basis for our findings and conclusions based on our audit objectives.

Background

PHMSA is responsible for setting and enforcing the federal minimum safety standards for gas and hazardous liquid pipelines, includingstandards for pipeline design, construction, operation, and maintenance.[5] As part of this

[3] GAO, Open Data: Treasury Could Better Align USAspending.gov with Key Practices and Search Requirements. GAO-19-72 (Washington, D.C.: Dec. 13, 2018). We developed these key practices by systematically evaluating and synthesizing information from literature on open data, as well as interviews with open data experts and good governance groups.

[4] We conducted interviews with nine stakeholder groups, including two federal agencies, one state association, three industry associations, one pipeline safety association, and two pipeline safety researchers.

[5] These standards apply to both interstate and intrastate pipelines. Interstate pipelines typically transport gas or hazardous liquid across state borders, while intrastate pipelines typically transport these products within a single state. States may assume regulatory, inspection, and enforcement responsibilities for intrastate pipelines, provided that they certify to PHMSA that

responsibility, PHMSA officials from offices in five regions and three districts periodically inspect interstate and some intrastate pipelines to oversee pipeline operators' compliance with federal requirements. PHMSA can issue enforcement actions when an inspector identifies probable violations of pipeline safety laws or regulations, or of a PHMSA order issued under those laws and regulations.[6] PHMSA may also issue enforcement actions in the course of investigating an identified safety issue or a pipeline accident.[7] According to PHMSA officials, the Agency's Office of Pipeline Safety has 170 authorized inspector positions whose responsibilities include inspecting 559 companies that operate about 377,000 miles of pipelines.

PHMSA has broad discretion in deciding what enforcement action, if any, to take against a particular pipeline operator to promote safety and ensure compliance with federal safety requirements.

- *Warning Letters* notify operators when PHMSA inspections or other oversight activities reveal less serious violations or program deficiencies.[8] Warning Letters direct the operator to correct the issues or be subject to potential future enforcement actions.
- *Notices of Amendment* allege inadequacies in the operator's plansand procedures to ensure safe operation of the pipeline, proposerevisions to the plans or procedures, and instruct the operator asto how to respond to the allegations.[9]
- *Notices of Probable Violation* allege the existence of one or more probable violations of pipeline safety laws, regulations, or related orders.[10] These notices are accompanied by either a proposed compliance order identifying the remedial actions PHMSA proposes

they have adopted and are enforcing the federal minimum safety standards. These states may also adopt additional or more stringent requirements, so long as they are compatible with federal requirements. States with certifications may also enter into agreements with PHMSA to participate in inspections of interstate pipelines as interstate "agents" of PHMSA. Interstate agents can identify and must report to PHMSA any violations or probable violations of the Pipeline Safety Laws or Regulations during inspections of interstate pipelines, but PHMSA retains enforcement authority over these pipelines.

[6] Regulations governing PHMSA's enforcement of these laws, regulations, and orders are located in 49 C.F.R. Part 190.

[7] PHMSA officials also set and enforce the federal minimum safety standards for and inspect liquefied natural gas and underground natural gas storage facilities. However, this report is focused on PHMSA's pipeline enforcement actions.

[8] See 49 C.F.R. § 190.205.

[9] *See* 49 C.F.R. § 190.206.

[10] *See* 49 C.F.R. § 190.207.

the operator to take or proposed civil monetary penalties, or both.[11] Notices of probable violation are the only typeof enforcement action that may include proposed civil monetary penalties. If PHMSA finds that an operator committed a violation, then it issues a Final Order, which includes the compliance order requiring the operator to take remedial action or the assessment of civil monetary penalties, or both, as applicable.[12]

Concerns about the transparency of PHMSA's enforcement actions against pipeline operators—including that the lack of transparency resulted in an incomplete accounting of the fines and penalties proposed and collected—led to the inclusion of a provision in the Pipeline Inspection, Protection, Enforcement, and Safety Act of 2006 (PIPES Act of 2006) to promote enforcement transparency.[13] Among other things, this provision required PHMSA to electronically post a monthly updated summary to the public of all of its open enforcement actions and include in each summary certain information such as the operator, alleged violation, and monetary penalties proposed and assessed.[14] In response to this requirement, PHMSA created its Enforcement Transparency website, which it launched in May 2007. The website provides public access to information on and documents from individual cases and various reports and data describing enforcement trends. The PIPES Act of 2020 later required PHMSA to make certain documents from each of its enforcement cases publicly available on its website, such as the operator's written responses to the allegations in the enforcement action.[15] According to PHMSA officials, the agency had been making these documents available on its website prior to the enactment of the PIPES Act of 2020.

[11] These enforcement actions must contain the options available to the operator for responding to the notice. The options include but are not limited to submitting written responses contesting the allegations, requesting mitigation or elimination of the proposed civil penalty, objecting to the compliance order, and requesting a hearing. Failure to respond constitutes a waiver of a right to contest the allegations.

[12] 49 C.F.R. § 190.213.

[13] Pub. L. No. 109-468, 120 Stat. 3486 (2006); H.R. Rep. 109-717, Part 2, at 8 (Dec. 5, 2006) (report of the House Committee on Energy and Commerce accompanying the PIPES Act of 2006).

[14] PIPES Act of 2006 § 6(a) (codified at 49 U.S.C. § 60135). According to the House Committee on Energy and Commerce, one purpose of this requirement was to provide timely and accurate information on pipeline safety enforcement cases to all interested parties. H.R. Rep. 109-717, Part 2, at 8.

[15] PIPES Act of 2020 § 108(a) (codified at 49 U.S.C. § 60117(b)).

Table 1. Key Practices and associated key actions for transparently reporting open government data

Key practice	Key actions
Engage with users	Identify data users and their needs
	Solicit and be responsive to user feedback
	Reach out to potential users to encourage data use
Fully describe the data	Disclose known data quality issues and limitations
	Disclose data sources and timeliness
	Clearly label the data and provide accompanying metadata
	Publish data under an open license and communicate licensing information to users
Provide free and unrestricted data	Do not charge users for access to the data
	Make government data open by default, while protecting sensitive or restrictedinformation
Provide data in useful formats	Provide users with detailed and disaggregated data
	Provide data downloads in a non-proprietary format
	Provide machine-readable data that can be downloaded in bulk and in selected subsets
	Make the data interoperable with other datasets
Facilitate data discovery for all users	Provide an interface that enables intuitive navigation and ensure that the most importantinformation is visible
	Provide users with appropriate interpretations of the data, such as visualizations orsummaries
	Ensure that the website's content is written in plain language
	Use central data repositories and catalogues to help users easily find the data they arelooking for
	Provide a search function that is optimized for easy and efficient use

Source: GAO. | GAO-23-105537.

Our prior work has identified key practices that can improve the transparency of such government websites and make data more accessible to users. Specifically, we have previously identified five key practices for transparently reporting open government data.[16] These key practices and associated key actions can assist managers of open government data programs, such as PHMSA, in the transparent presentation of their data and can help users find the information they need. See table 1 for the full list of key practices and associated actions.

[16] GAO-19-72. Open government data are information that can be freely used or modified or shared by anyone for any purpose.

PHMSA Has a Standardized Process to Record Enforcement Information and Makes Correspondence Publicly Available

PHMSA Records Pipeline Enforcement Information Using a Standardized Process

PHMSA's regional and headquarters staff have a standardized process to record and maintain enforcement information in an electronic case file for pipeline safety enforcement cases. Specifically, throughout an enforcement case, PHMSA regional staff conduct inspections and investigations, collect evidence, and correspond with pipeline operators. To record the information they collect, regional staff upload documents and other information to an electronic case file. Within headquarters, PHMSA's Enforcement Division is responsible for maintaining this information in the case file and validating and correcting, if needed, the enforcement information. According to officials from this division, they also provide support to PHMSA regional staff as they enter information into the system. PHMSA attorneys in headquarters also draft some enforcement documents for the case file. See figure 1 for an example of PHMSA's process to record enforcement information for a case involving a notice of probable violation.[17]

PHMSA's regulations specify what documents officials should include in the case file, which varies based on the type of enforcement action.[18] Generally, the case file should include correspondence between PHMSA regional officials, PHMSA's Office of Chief Counsel, and operators. The file also is to include other documents such as violation reports, including evidence such as maps or drawings, and additional documents needed for evidence. According to PHMSA guidance, draft documents and documents not needed to adequately document the case are not to be included in the case file. PHMSA officials reported that this includes itemssuch as inspection checklists, inspector notes, and draft documents.

---

[17] Other types of enforcement actions have different steps in PHMSA's standardized process. For example, PHMSA does not conduct hearings or issue Final Orders for pipeline enforcement cases that include only Warning Letters. *See* 49 C.F.R. § 190.205.

[18] See 49 C.F.R. § 190.209.

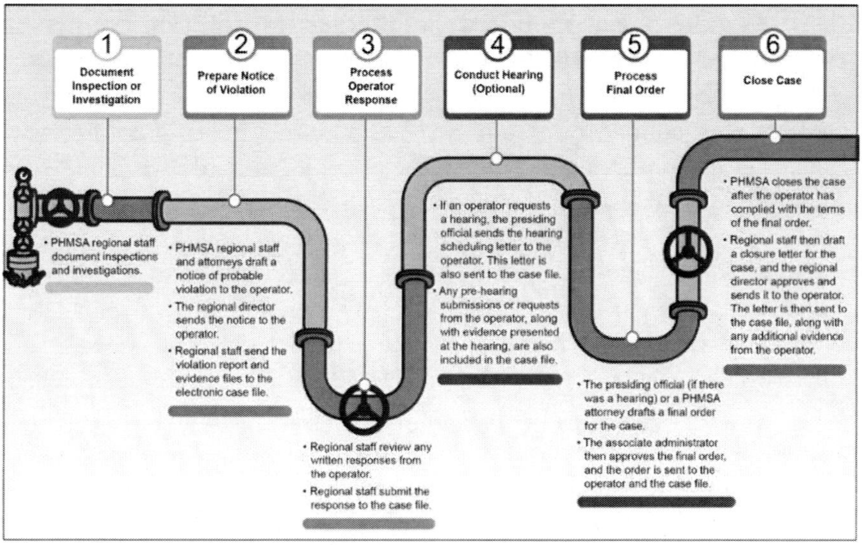

Source: GAO analysis of PHMSA guidance. | GAO-23-105537.

Note: Other types of enforcement actions require different steps in PHMSA's standardized process.

Figure 1. The Pipeline and Hazardous Materials Safety Administration's (PHMSA) Process to Record Enforcement Informationfor a Case Involving a Notice of Probable Violation.

Since 2020, PHMSA has recorded and maintained these case files through a centralized electronic system called the Work Management System. The Work Management System allows PHMSA regional officials to upload enforcement documents directly to the system. Under PHMSA's previous system, regional officials would email documents to a coordinator in headquarters, who would then upload them. PHMSA headquarters officials said they are continuing to improve the new Work Management System to address technical issues and increase the ease of use for regional staff. For example, officials reported that they are working with the developer to fix bugs that can result in missing data.

PHMSA officials said they have a development plan to fix these issues and are taking several steps to ensure data accuracy and reliability as they work to resolve technical issues in the system. For example, PHMSA holds daily quality control meetings to review the previous day's data entries and runs monthly quality control scripts to detect data errors.

PHMSA Makes Correspondence and Other Information Publicly Available on Its Website

PHMSA is required by statute to make certain types of enforcement documents and information from the electronic case file available to the public. PHMSA does so by uploading such documents to its Enforcement Transparency website.[19] Generally, these documents include correspondence between PHMSA and the operator, such as notice lettersand the operator's written response, if any. PHMSA also is required to upload other correspondence that describes the outcome of the enforcement case (e.g., Final Order, which includes details such as PHMSA's final decision on whether there was a violation and the assessed civil penalty amount or compliance order, or both, as applicable.) According to its guidance, PHMSA also uploads documents beyond what is required, specifically case closure letters and post-hearingsubmissions from operators to the website. Finally, PHMSA officials reported that in addition to these case documents, they also upload referrals to the Department of Justice and court enforcement actions along with documents explaining the circumstances related to unpaid penalty cases, such as bankruptcy.

PHMSA generally does not make the evidence in the case file public on its Enforcement Transparency website and is not statutorily required to do so. This evidence includes items such as the violation report, along with maps or drawings. PHMSA officials reported that these documents can contain proprietary or security-restricted information and that consulting with operators to redact each document would be resource intensive andslow the enforcement process. Officials noted, however, that members ofthe public could request these documents through a Freedom of Information Act request; in which case, PHMSA would review and redactthe documents as needed.[20] See table 2 for a list of enforcement documents that could be contained in the electronic case file, and which of those documents PHMSA is required to make public on its enforcement website.

[19] Under the Freedom of Information Act (FOIA), codified as amended at 5 U.S.C. § 552, federal agencies are generally required to disclose any information requested under FOIA, unless it falls under one of nine exemptions.

[20] Under the Freedom of Information Act (FOIA), codified as amended at 5 U.S.C. § 552, federal agencies are generally required to disclose any information requested under FOIA, unless it falls under one of nine exemptions.

Table 2. List of documents from enforcement case files that the pipeline and hazardous materials safety administration(PHMSA) is legally Required to Make Public and Makes Available on Its Website

Type of enforcement documents included in case file	Legally required to makepublic[a]	Available on PHMSA'swebsite[b]
Correspondence from PHMSA		
Notice Letter	Yes	Yes
Warning Letter	Yes	Yes
Letter of Concern	No	No
Request for Specific Information letter	No	No
Withdrawal of Notice Letter or Warning Letter	Yes	Yes
Withdrawal of Letter of Concern	No	No
Response to operator's request for hearing	No	No
Response to operator's request for time extension	No	No
Region recommendation (for final action based on operator's writtenresponses)	No	No
Final Order (and amendments)	Yes	Yes
Corrective Action Order (and amendments)[c]	Yes	Yes
Decision on Corrective Action Order	Yes	Yes
Order Directing Amendment of Operating Plans and Procedures (andamendments)	Yes	Yes
Decision on Petition for Reconsideration	Yes	Yes
Safety Order (and amendments)	Yes	Yes
Consent Agreement and Order	Yes	Yes
Closure Letter	No	Yes
Hearing Scheduling Letter	Yes	Yes
Correspondence from Operator		
Response to Notice Letter	Yes	Yes
Response to a Warning Letter or Letter of Concern	No	No
Response to Request for Specific Information letter	No	No
Operator's amended operating plans and procedures	No	No
Request for time extension	No	No
Request for hearing	Yes	Yes
Post-hearing submissions	No	Yes
Response to an Order Directing Amendment	No	No
Response to a Final Order	No	No
Response to a Corrective Action Order	No	No
Response to a Safety Order	No	No
Petition for Reconsideration	Yes	Yes
Other Documents		
Violation Report (including evidence such as maps, drawings,respondent record forms, etc.)	No	No
Additional documents needed to document the case or evidence	No	No

Source: GAO analysis of PHMSA guidance. | GAO-23-105537.

[a] These legal requirements only apply to PHMSA. See Protecting Our Infrastructure of Pipelines and Enhancing Safety Act of 2020, Pub. L. 116–260, div. R, § 108(a), 134 Stat. 1182, 2221 (2020) (codified at 49 U.S.C. § 60117(b)); Pipeline Inspection, Protection, Enforcement and Safety Act of 2006, Pub. L. No. 109-468, § 6(a), 120 Stat. 3486, 3491 (2006) (codified at 49 U.S.C. § 60135). We did not include other legal requirements to make certain agency records and information publicly available that apply more broadly to all federal agencies. See, e.g., Freedom of Information Act, 5U.S.C. § 552.

[b] These are the documents that PHMSA is to upload to its website per PHMSA guidance. This guidance also outlines its quality assurance process to ensure the documents are posted as required. GAO confirmed that PHMSA posted at least one of each of these types of documents, including those that it is statutorily required to make public. We did not verify whether PHMSA has posted every document that it is legally required to post for each enforcement case on its Enforcement Transparency website.

[c] Decisions on Corrective Action Orders are only issued in response to an operator contesting an order and affirms the need for the order, with or without amendments, or withdraws the order. See 49 C.F.R. § 190.233(c)(5).

In addition to these documents from the case file, PHMSA makes enforcement information available online in other forms. For example, PHMSA creates a summary page for each case, which displays selected information contained in the enforcement documents posted on the website. This summary page includes details such as the operator; date the case was opened, case status (open or closed); and the proposed, assessed and collected civil penalties (if any.) Additionally, PHMSA has an electronic reading room, which provides links to Final Orders and otherenforcement documents, along with frequently requested enforcement records.[21] PHMSA also posts information that describes its enforcement processes on its website. For example. PHMSA posts a "Civil Penalty Summary" that describes factors PHMSA considers when setting a proposed civil penalty, and guidance documents that describe PHMSA'senforcement process.

[21] PHMSA created this electronic reading room in order to comply with FOIA requirements.

PHMSA's Pipeline Enforcement Website Does Not Fully Align with Key Practices for Transparently Reporting Data

In a 2018 report, we identified five key practices for transparently reporting on open government data, including 18 specific actions agencies should take to align with these practices.[22] Open data can foster accountability and public trust by providing citizens with information on government activities and their outcomes.

Our analysis of PHMSA's Enforcement Transparency website found it met 14 of the 18 specific actions, resulting in the website fully aligning with two, and partially aligning with three, of the five key practices for transparently reporting on open government data. Specifically, we found that PHMSA fully engages with users and fully describes its data, but does not fully (1) provide free and unrestricted data, (2) provide data in useful formats, or (3) facilitate data discovery for all users. Figure 2 provides our assessment of the alignment of PHMSA's Enforcement Transparency website with each key practice and related actions.

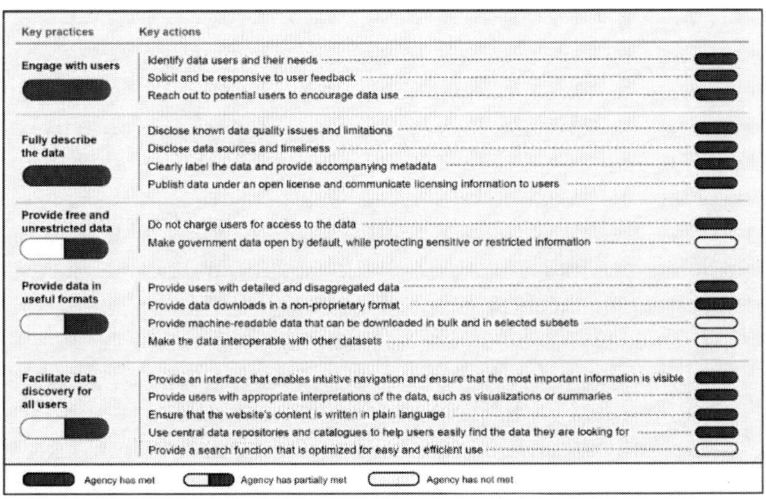

Source: GAO analysis of PHMSA's Enforcement Transparency website and key practice. | GAO-23-105537.

Figure 2. Alignment of the Pipeline and Hazardous Materials Safety Administration's (PHMSA) Enforcement TransparencyWebsite with Key Practices and Actions for Transparently Reporting on Open Government Data.

---

[22] GAO-19-72.

Key Practice 1: PHMSA Engages with Users to Identify Needs and Solicit Feedback.

Following the key practice of engaging with users to identify their needs and solicit feedback helps to ensure that their needs are met. We found *PHMSA met all three actions associated with this key practice.*

- *Action: identify data users and their needs.* PHMSA officials stated that users of the Enforcement Transparency website include operators, state partners, members of Congress and their staff, the public, emergency responders, local officials, and other PHMSA staff. According to these officials, PHMSA discusses the use of the website with these stakeholders and receives feedback to identify users' needs through meetings, conferences, and other communications. PHMSA officials also reported that they track the number of hits they receive on the Enforcement Transparency website. By identifying who is using the data and what content or features are important to them, data providers can better prioritize their efforts to present information to data consumers.
- *Action: solicit and be responsive to user feedback.* PHMSA provides a hyperlink on the Enforcement Transparency website that redirects users to a general contact page where they can submit feedback or questions about the website. PHMSA officials stated that they consider the feedback in considering changes to improve the website; for example, staff began posting hearing schedules in 2019 because of inquiries received through the website. Officials also explained that they receive feedback during conferences with stakeholders. Soliciting and being responsive to user feedback can help ensure that the website meets users' needs.
- *Action: reach out to potential users to encourage data use.* PHMSA officials stated that they reach out to potential users of the Enforcement Transparency website through conversations with stakeholders and in meetings with members of Congress. For example, PHMSA officials reported that they presented an overview of the Enforcement Transparency website at various industry conferences in the past. Actively engaging potential users can provide an opportunity to educate them on how the data can be appropriately used and encourage innovation.

Key Practice 2: PHMSA Provides Information on its Enforcement Data

The key practice of providing information about a dataset allows users to determine whether it is suitable for their intended purpose, and make informed decisions about whether and how to use it. We found that PHMSA met all four key actions associated with this key practice.

- *Action: disclose known data quality issues and limitations.* PHMSA officials described the data entry procedures and quality control processes they use to ensure the information collected by inspectors is accurate and complete. Further, officials stated that they continue to review the Work Management System processes to identify any potential data quality issues. In the meantime, they have implemented additional data procedures and quality control measures and checks to ensure potential issues would not affect the enforcement data. As a result, PHMSA officials reported that there are no known data quality issues or limitations to disclose on the Enforcement Transparency website.[23]
- *Action: disclose data sources and timeliness.* PHMSA's Enforcement Transparency Website includes a source line that states that the data published are from its Safety Monitoring and Reporting Tool and the Work Management System, and that they are updated monthly as additional cases are initiated and other cases are resolved.[24] PHMSA officials told us that this generally occurs in the first two weeks of each month. Disclosing where the data come from and how frequently they are updated provides context that helps users judge their quality and determine whether they can be appropriately used for the intended purpose.
- *Action: clearly label the data and provide accompanying metadata.* PHMSA labels all of the data in tables and summaries on the Enforcement Transparency Website and provides descriptions of each of these. Additionally, the Department of Transportation (DOT) publishes a data inventory website that includes metadata for PHMSA's Enforcement Transparency website. (See fig. 3.) Metadata are structured information that describe, explain, locate, or otherwise make it easier to retrieve, use, or manage datasets. Metadata can include attributes such as the date the data were created or modified,

[23] Additionally, we reviewed PHMSA's data validation processes and guidance and also found them to be sufficiently reliable.

[24] Prior to the implementation of the Work Management System, PHMSA used the Safety Monitoring and Reporting Tool to maintain records of each enforcement action's case file.

the license used, or the name and contact information for the owner of the dataset. Clearly labeling data allows users to easily find information about a dataset, and making metadata machine readable greatly increases their utility.

- *Action: publish data under an open license and communicate licensing information to users.* PHMSA publishes data to the Enforcement Transparency website under an open license, which means that there are few to no restrictions on how the data can be used. The data inventory website published by DOT provides information on the license that applies to PHMSA's enforcement data. Documenting licensing information for a dataset provides users with information about how they may use the data, including whether there are any restrictions, such as copyrights.

About this Dataset

Views 1 Downloads 0

Data Provided by Pipeline and Hazardous Materials Safety Administration

Dataset Owner Daniel Morgan

Contact Dataset Owner

Common Core

Publisher	Pipeline and Hazardous Materials Safety Administration
Contact Name	Seong Hwang
Contact Email	seong.hwang@dot.gov
Bureau Code	021:50
Program Code	021:044
Public Access Level	public
Geographic Coverage	United States
Temporal Applicability	2002-01-01/P1M
Theme	transportation
Data Dictionary	https://primis.phmsa.dot.gov/comm/EnforcementProgram.htm
Is Quality Data	true
Language	en-US
Primary IT Investment UII	021-500121022
Homepage	https://primis.phmsa.dot.gov/comm/reports/enforce/enforcement.html
Issued	2002-01-01
License	http://www.usa.gov/publicdomain/label/1.0/
Update Frequency	R/P1M

Source: DOT's Open Data Catalog. | GAO-23-105537.

Figure 3. Metadata for the Pipeline and Hazardous Materials Safety Administration's (PHMSA) Enforcement Transparency Website Provided on the Department of Transportation's (DOT) Open Data Catalog.

Key Practice 3: PHMSA Provides Enforcement Data for Free but Does Not Publish All Data Collected.

Under this key practice, open data should be freely and equally available to users without restrictions. We found that PHMSA met the first action associated with this key practice but did not meet the second.

- *Action: do not charge users for access to the data.* PHMSA's Enforcement Transparency Website and the data it contains on enforcement activities are available to view or download by the public at no cost. Additionally, users of the website are not required to register for an account to view or download the data and case information. Providing data for free can help ensure equitable access to users regardless of their ability to pay.
- *Action: make government data open by default while protecting sensitive or restricted information.* PHMSA collects and records data on over 200 variables that describe each enforcement case. However, the data are not open by default as not all variables are published in the dataset on the Enforcement Transparency website. For example, PHMSA collects and records variables that indicate whether the case involves intrastate or interstate pipelines and onshore or offshore pipelines, as well as variables that describe the proposed violations and relevant regulations. These variables are not all sensitive or restricted (the information can be found in the documents published within each case file). However, PHMSA does not include them in the enforcement dataset.

  Key practices suggest that making government data open by default can minimize the burden on agencies of responding to information requests. Additionally, providing complete, open data ensures the data are equally open to all types of users. In contrast, when government information is available by request, it may favor citizens with greater information about and access to government institutions. For example, one stakeholder noted that to use the enforcement data for analyses, a team member must go in to each case file to find the regulation that was violated and other important data fields, which requires them to read many documents included in each case. However, members of the public may not have the ability or knowledge of how to access this information. The stakeholder also noted that additional variables could be useful, such as the regulation that was violated and the incident that occurred, which would allow users to retrieve the information they need from a complete dataset without locating and reviewing the documents for each case.

Key Practice 4: PHMSA's Enforcement Data Are Detailed and in Standardized Formats but Lack Full Usability.

This key practice notes that data are most useful when they are provided in formats that allow a variety of analyses. We found that PHMSA met two of the four actions associated with this key practice.

- *Action: provide users with detailed and disaggregated data.* PHMSA allows users to view and download disaggregated enforcement data on all enforcement activities. Disaggregated data are those that are provided in as much granularity as possible. PHMSA's enforcement data give users information at anindividual case level, with variables for the case number assigned by PHMSA; the pipeline operator's name; the PHMSA region overseeing the case; the case status (open or closed); and relevant dates. Additionally, Notice of Probable Violation data detail the amount of civil penalties proposed. Users can view and download these case-level data one year at a time. See figure 4 for an example of disaggregated data at the enforcement case level displayed on the Enforcement Transparency website. Key practices suggest that data are most useful when they are provided in a disaggregated format so users can analyze the datain a variety of ways.

Listing of Cases Initiated

Nationwide

This report lists all of the enforcement cases initiated by PHMSA in a given year, beginning in 2002. The yearly tables include the date the case was opened, the name of the operator involved, the PHMSA Region initiating the enforcement action(s), the corresponding date on which the case was closed (if applicable), the current status of the case, and PHMSA's unique case number. Separate reports provide a listing of cases based on the type of enforcement action involved.

2022 YTD | 2021 | 2020 | 2019 | 2018 | 2017 | 2016 | 2015 | 2014 | 2013 | 2012 | 2011 | 2010 | 2009 | 2008

Cases Initiated: 2020 [1]

Date Opened	Operator	Region	Date Closed	Status	Case Number
01/02/20	NATURAL GAS PIPELINE CO OF AMERICA (KMI)	Eastern	02/07/20	CLOSED	120201006M
01/02/20	COLUMBIA GAS TRANSMISSION, LLC	Eastern	07/16/20	CLOSED	120201005M
01/02/20	TEXAS EASTERN TRANSMISSION, LP (SPECTRA ENERGY PARTNERS, LP)	Eastern	01/02/20	CLOSED	120201001W
01/02/20	ANR PIPELINE CO	Eastern	01/02/20	CLOSED	120201004W
01/02/20	COLUMBIA GAS TRANSMISSION, LLC	Eastern	01/02/20	CLOSED	120201003W
01/02/20	DTE GAS COMPANY	Eastern	01/02/20	CLOSED	120201002W
01/08/20	VALERO PARTNERS OPERATING CO. LLC	Southwest	10/01/20	CLOSED	420205001
01/08/20	CHENIERE CORPUS CHRISTI PIPELINE	Southwest	02/10/20	CLOSED	420201003M
01/09/20	WILLIAMS OIL GATHERING, LLC	Southwest	01/09/20	CLOSED	420207001W
01/13/20	CHEVRON PIPE LINE CO	Southwest	01/13/20	CLOSED	420205003W
01/13/20	WILLIAMS OIL GATHERING, LLC	Southwest	02/24/20	CLOSED	420205002M
01/13/20	NATURAL GAS PIPELINE CO OF AMERICA (KMI)	Eastern	01/13/20	CLOSED	120201007W
01/14/20	CHEVRON PIPE LINE CO	Southwest	03/05/20	CLOSED	420205004M
01/21/20	BUCKEYE PARTNERS, LP	Eastern	04/10/20	CLOSED	120206001
01/22/20	ENABLE GAS TRANSMISSION, LLC	Southwest	02/01/21	CLOSED	420201004
01/23/20	GULFSTREAM MANAGEMENT & OPERATING SERVICES,LLC	Southern	04/09/20	CLOSED	220201001
01/27/20	FREEPORT LNG DEVELOPMENT, L.P.	Southwest	04/15/20	CLOSED	420201001M
01/27/20	DTE GAS COMPANY	Central	08/04/20	CLOSED	320201002M

Source: PHMSA'sEnforcement Transparency website. | GAO-23-105537.

Figure 4. Selected Enforcement Cases Initiated by the Pipeline and Hazardous Materials Safety Administration (PHMSA).

- *Action: provide data downloads in a non-proprietary format.* PHMSA provides data on the Enforcement Transparency website in a non-proprietary format that does not require a specific commercial software to access. By selecting "export table" at the bottom of each table, users are able to download the data and open the file using any software available to them. To ensure broad and equitable access, data downloads should be available in formats that do not exclude users that do not have access to specific commercial software.

We found that PHMSA did not meet two of the actions associated with this key practice.

- *Action: provide machine-readable data that can be downloaded in bulk and in selected subsets.* Users of PHMSA's Enforcement Transparency website cannot download a bulk dataset for all cases opened by PHMSA from 2002 (the first year of data available on the website) through the current year. Additionally, users are unable to obtain a subset of the data by filtering out cases with certain criteria or in a certain time frame. Instead, the data are available for download as summaries (e.g., total number of enforcement cases initiated each year) or disaggregated at the case level for a single year or single enforcement type. As a result, we were unable to download a bulk dataset of enforcement cases from 2010 to 2020 to perform our own analyses of the enforcement data. (See app. 1.) When askedif there were a way to make a bulk dataset available for download,which would also allow users to access subsets of the data, PHMSA officials stated that this was feasible. However, as of September 2022, a bulk dataset is not available for download on PHMSA's Enforcement Transparency website. Making data available to download in bulk allows users who need the full dataset or selected subsets to access that information easily, rather than retrieving it record by record. For example, PHMSA could make a bulk dataset available in a file format such as a comma-separated values file that can be opened using common office software or through an application-programming interface.
- *Action: make the data interoperable with other datasets.* Users cannot easily match records on PHMSA's Enforcement Transparency website with other relevant PHMSA datasets, such as the accident-incident dataset because there is no shared variable between the

datasets that would allow them to do so. PHMSA may pursue an enforcement action in response to a pipeline accident. According to PHMSA officials, users could match these enforcement records to the accident dataset by using the operator search function on the Enforcement Transparency Website. To do so, a user could download the accident dataset and identify the operator concerned and the date the accident occurred. Concurrently, the user could use the operator search function to identify any enforcement actions taken against that operator in the same year to attempt to match the records.

However, if an operator had multiple enforcement actions taken against it or was involved in multiple accidents in the same year, the user may not be able to identify which cases are related to which accidents, as there is not a unique identifier variable common to both datasets. Additionally, it may take PHMSA several months to open an enforcement case after an accident and the resulting investigation occur (e.g., the accident could happen in 2020 and the enforcement case could be initiated in 2021), a lag that could make it more difficult for users attempting to match records between the two datasets. Making the enforcement data interoperable with other datasets, for example by including a unique identifier variable that would allow users to match enforcement case data with accident data, could make them more useful by creating new opportunities for analysis by linking datasets together. For example, one stakeholder we spokewith stated that connecting the enforcement and accident datasetswould allow users to understand if an accident resulted in PHMSAtaking an enforcement action.

Key Practice 5: PHMSA Makes Enforcement Data Easy to Understand but Does Not Fully Facilitate Data Discovery

According to this practice, data discovery is facilitated by features that enable users to easily explore the data. We found that PHMSA met fourof the five actions associated with this key practice.

- *Action: provide an interface that enables intuitive navigation and ensure that the most important information is visible.* PHMSA's Enforcement Transparency website displays an introductory page that provides users with a description of the various ways the enforcement data can be viewed, followed by links to each page. These pages include an overall summary of PHMSA's enforcement activity and breakdowns by enforcement action, enforcement case status, and civil penalty status. Ensuring that the data are provided on a website that is simple and intuitive facilitates data discovery for all users.

- *Action: provide users with appropriate interpretations of the data, such as visualizations and summaries.* PHMSA publishes national summaries of its enforcement activity data. For example, users can view a summary of the total number of cases opened and closed each year for all PHMSA regions, and a breakdown, by year, of the number of cases opened for each enforcement action type. Such summaries can help users explore the data.
- *Action: ensure that the website's content is written in plain language.* PHMSA defines enforcement action terms and enforcement topics discussed throughout the Enforcement Transparency website, which provides users with clear and direct language to aid them in understanding the information and data published on the website. Additionally, PHMSA provides a glossary in the sidebar of every page that gives users an overview of basic pipeline and enforcement terms and technical phrases. In cases where it is necessary to use technical language, providing a glossary and defining key terms can help make the information more understandable to users.
- *Action: use central data repositories and catalogues to help users easily find the data they are looking for.* PHMSA publishes a link to the Enforcement Transparency Website on DATA.gov, a central data catalogue for a variety of U.S. government datasets. PHMSA also includes a description of the enforcement dataset that can help users find data suitable for their purposes. Linking to the enforcement data from centralized data repositories and catalogues such as DATA.gov can facilitate discovery of the enforcement data for users who are interested in pipeline safety but unaware of PHMSA's Enforcement Transparency website.

We found that PHMSA did not meet one action associated with this key practice.

- *Action: provide a search function that is optimized for easy and efficient use.* PHMSA currently allows users to sort the enforcement data by operator name (see fig. 5), but users are unable to search with other key terms or variables. Having such asearch function would allow users to find information that is relevant to their interests. For example, one stakeholder noted that a search function that would allow users to search with a specific variable, such as by the regulation violated, would be useful in analyzing civil monetary penalties and outcomes for specific violations. Another stakeholder noted that the ability to search by the regulation violated would allow users to identify all enforcement actions for the particular safety issue with which theyare concerned.

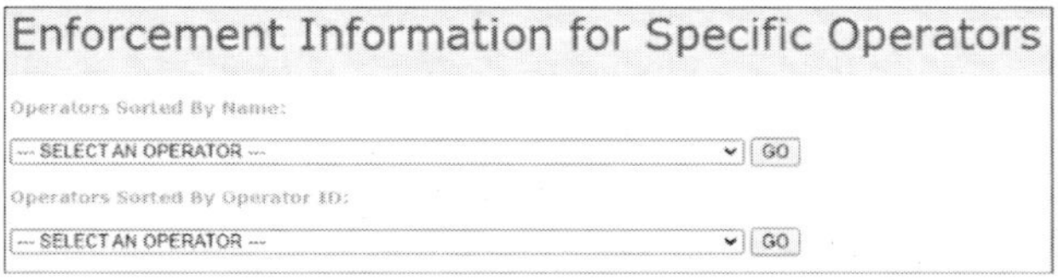

Source: PHMSA's Enforcement Transparency website. | GAO-23-105537.

Figure 5. The Pipeline and Hazardous Materials Safety Administration's (PHMSA) Operator Search Function on theEnforcement Transparency Website.

In discussions with PHMSA officials regarding why the Enforcement Transparency website does not fully align with key practices, they told us it is because PHMSA has limited staff and they are focusing their efforts on addressing issues with the Work Management System. They hope to complete these improvements to the system by the end of 2022. They noted that PHMSA is considering ways to make the Enforcement Transparency website more useful, and acknowledged that they would like to move to a more dynamic website that would facilitate better searching and data views. Nevertheless, they told us they remain focused on other priorities first.

However, aligning the Enforcement Transparency website with all of the key practices would be beneficial. For instance, after the Work Management System is stabilized, a focused effort by PHMSA to update the website would improve the usability and transparency of its enforcement information. If users had access to a complete enforcement data file, they could more efficiently search PHMSA records by filtering data instead of searching individual records. Additionally, a more robust search function would enable users to more easily identify specific cases for their own purposes, such as looking up the enforcement history of pipelines located near their home.

Conclusion

Although hazardous liquid pipeline accidents are relatively infrequent, leaks and ruptures still occur and can cause fatalities, injuries, and environmental damage. Accordingly, public access to pipeline enforcement information is important for fostering accountability and trust regarding the federal government's pipeline safety efforts. PHMSA has taken steps to increase public access to pipeline enforcement information through the creation of its Enforcement Transparency website. The website meets two of the key practices and most of the key actions for transparently reporting data, such as providing the data on the website for free, engaging the public online and in person, and providing disaggregated data for download.

However, PHMSA could better align its Enforcement Transparency website with other key practices and associated actions, to improve the usefulness and transparency of the website. Specifically, enabling the complete pipeline enforcement dataset to be downloaded in bulk could better enable users to analyze data from multiple years or filter out cases not needed for analyses. In addition, connecting PHMSA's accident- incident and enforcement datasets, for example by adding a unique identifier variable, would help users track a case from pipeline accident through enforcement action. Moreover, although the website currently allows filtering by operator, permitting the enforcement data to be searched by other variables—such as the regulation violation or pipeline type—could better facilitate data analysis and trend identification by the public.

Recommendations for Executive Action

We are making three recommendations to PHMSA. Specifically:

- The Administrator of PHMSA should update PHMSA's Enforcement
- Transparency website to provide complete, machine-readable data that can be downloaded in bulk and in selected subsets. (Recommendation 1)
- The Administrator of PHMSA should update PHMSA's enforcement data to be interoperable with other relevant datasets, such as PHMSA's accident-incident dataset (e.g., by adding a unique identifier variable.) (Recommendation 2)
- The Administrator of PHMSA should update PHMSA's Enforcement Transparency website to provide a search function that allows users to find relevant information, which could include the ability to search by theregulation violated. (Recommendation 3)

Agency Comments

We provided DOT with a draft of this report for review and comment. In written comments, reproduced in appendix II, DOT concurred with our recommendations.

We are sending copies of this report to the appropriate congressional committees, the Secretary of Transportation, and other interested parties.In addition, the report is available at no charge on the GAO website at https://www.gao.gov.

If you or your staff have any questions about this report, please contact me at (202) 512-2834 or repkoe@gao.gov. Contact points for our Officesof Congressional Relations and Public Affairs may be found on the last page of this report. GAO staff who made key contributions to this report are listed in appendix III.

Elizabeth Repko
Director, Physical Infrastructure Issues

Appendix I: Analysis of Pipeline and Hazardous Materials Safety Administration's (PHMSA) Data on Pipeline Enforcement Actions

The Protecting our Infrastructure of Pipelines and Enhancing Safety Act of 2020 (PIPES Act of 2020) includes a provision for GAO to review information on pipeline enforcement actions that PHMSA makes publicly available on the internet.[25] We analyzed PHMSA enforcement data to describe the number and type of enforcement cases that were initiated from 2010 to 2020.

[25] Pub. L. No. 116-260, div. R, § 108(a), 134 Stat. 1182, 2221 (2020) (codified at 49 U.S.C.§ 60117(b)). PHMSA's general authority is under the Pipeline Safety Laws codified at 49U.S.C. § 60101 et seq., and its regulations are located in 49 C.F.R. Parts 190-199. The regulations governing PHSMA's enforcement of these laws, regulations, and its orders issued under them are specifically located in 49 C.F.R. Part 190.

Methodology

We analyzed enforcement data that PHMSA provided directly to us, rather than data publicly available on the website, because it was more efficient to do so. PHMSA's publicly available data are not available in bulk downloads and cannot be filtered to remove certain cases, such as those dealing with liquefied natural gas (LNG) facilities, which were not relevant to our analysis.

PHMSA provided a file with 35 variables that we selected, one of which was an indicator for an LNG case. Some of these variables have information that would be included in a case file but would not be listed as a value in a data table on the enforcement website. These include variables that indicate whether the case dealt with intrastate or interstate pipelines, onshore or offshore pipelines, and hazardous liquid or gas pipelines or LNG facilities, as well as an indicator for the type of enforcement action (e.g., Corrective Action Order, Notice of Probable Violation), the alleged violations and regulations involved, and the dates PHMSA began and ended the inspection prior to pursuing action.

To ensure that the data were sufficiently reliable for the purposes of this analysis, we reviewed PHMSA's Work Management System data dictionary and user guide to familiarize ourselves with the enforcement activity data. We then reviewed the processes PHMSA described in its data validation report and in interviews. We also manually tested the data for obvious errors, such as inconsistent dates and missing values in key columns. We identified no limitations with the data PHMSA provided and found it to be sufficiently reliable for our purposes.

Analysis

To conduct this analysis, we summarized the cases PHMSA initiated fromthe years 2010 to 2020. The summaries displayed in the following table and figures include enforcement actions that dealt with both interstate and intrastate, onshore and offshore, and gas and hazardous liquid pipelines throughout PHMSA's five regions. These regions (see fig. 5)—the Central, Eastern, Southern, Southwest, and Western—conduct inspections and investigations, collect evidence, and correspond with operators during enforcement cases. PHMSA assigns each pipeline operator to one of the five regions based in part on the location of the operators' pipelines within the boundaries of the regions. Many interstate operators' pipelines are located in multiple regions.

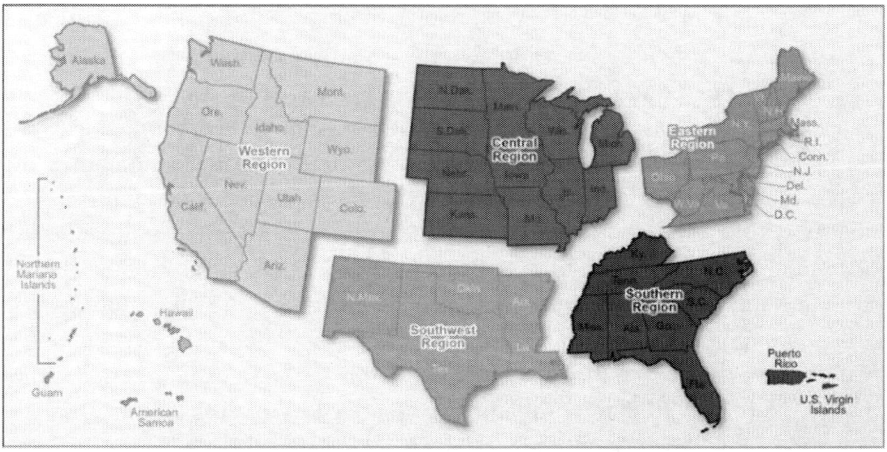

Source: GAO depiction of PHMSA information. | GAO-23-105537.

Figure 6. Map of the Pipeline and Hazardous Materials Safety Administration (PHMSA) Regions.

Table 3 displays a count of total cases initiated by region. The number of enforcement cases initiated by PHMSA fluctuated for each region over the 10-year period.

Figure 7 displays a count of total cases initiated by region, by enforcement action type. Generally, PHMSA relied on Notices of Amendment, Notices of Probable Violations, and Warning Letters whenpursuing enforcement actions throughout the five regions.

Table 3. Number of Cases the Pipeline and Hazardous Materials Safety Administration (PHMSA) Initiated by Region, from 2010to 2020

Year	Central	Eastern	South- ern	South- west	West- ern	Total Cases per Year
2010	22	26	23	27	55	153
2011	32	31	42	40	69	214
2012	43	57	48	56	82	286
2013	54	81	43	49	45	272
2014	19	21	34	49	36	159
2015	22	64	16	40	59	201
2016	12	33	15	52	62	174
2017	29	57	14	57	99	256

Year	Central	Eastern	South-ern	South-west	West-ern	Total Cases per Year
2018	23	64	18	43	49	197
2019	41	56	18	41	62	218
2020	48	61	12	50	24	195
Total Cases per Region	345	551	283	504	642	2325

Source: GAO analysis of PHMSA enforcement data. | GAO-23-105537.

Note: For this analysis, we excluded cases that dealt with liquefied natural gas (LNG) facilities. The Protecting our Infrastructure of Pipelines and Enhancing Safety Act of 2020 asked us to look specificallyat pipeline enforcement actions. Pub. L. No. 116-260, div. R § 108(a), 134 Stat. 1182, 2221 (2020) (codified at 49 U.S.C. § 60117(b)). LNG is stored in facilities connected to natural gas pipelines but is not transported through them due to its highly volatile nature. We also excluded cases that resulted in an Excavation Notice of Probable Violation from the scope of this analysis. These enforcement casesdo not deal with the regulations and safety requirements for operating a pipeline and instead focus on companies that have damaged pipelines while excavating. Additionally, we excluded cases under thepurview of the headquarters. In the data file PHMSA provided, there are 50 Warning Letter cases that were opened and closed by the headquarters office on September 7, 2010.

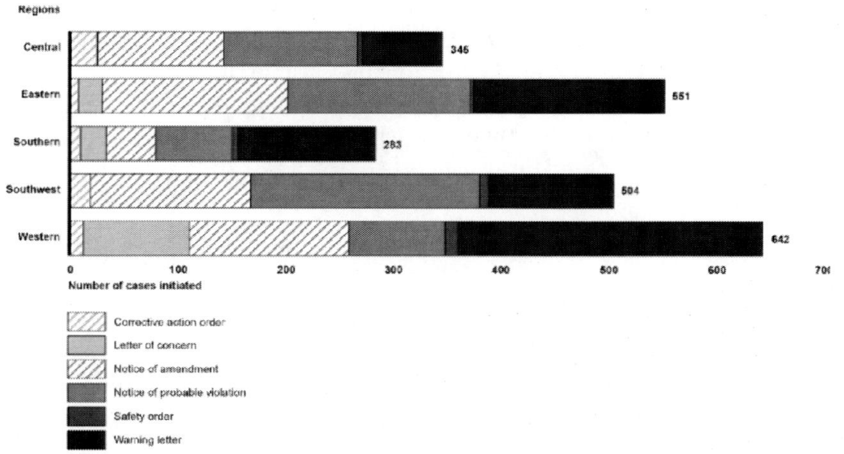

Source: GAO analysis of PHMSA enforcement data. | GAO-23-105537.

Note: For this analysis, we excluded cases that involved liquefied natural gas (LNG) facilities and excavators that did not comply with PHMSA's requirements for protecting underground pipelines from excavation-related damage; these requirements are located in 49 C.F.R. Part 196. Additionally, we excluded cases under the purview of the headquarters.

Figure 7. Total Number of Cases the Pipeline and Hazardous Materials Safety Administration (PHMSA) Initiated by Region andEnforcement Action Type, from 2010 to 2020.

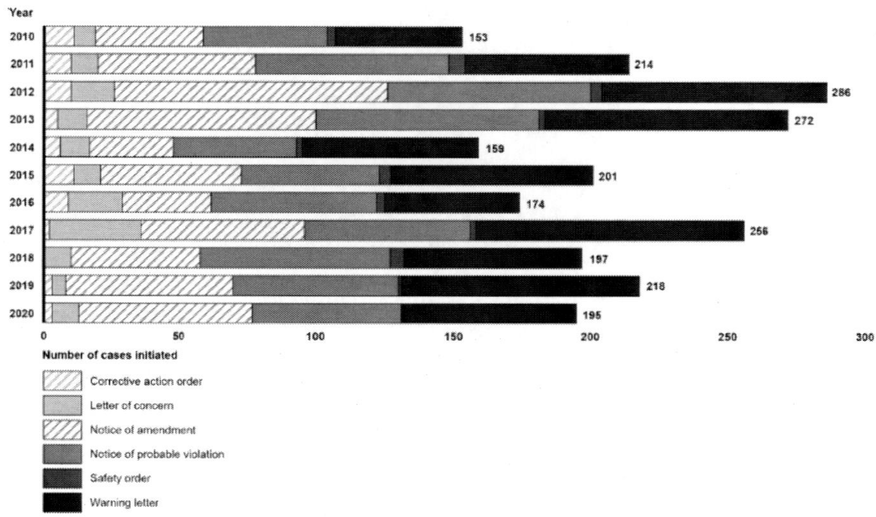

Source: GAO analysis of PHMSA enforcement data. | GAO-23-105537.

Note: For this analysis, we excluded cases that involved liquefied natural gas (LNG) facilities and excavators that did not comply with PHMSA's requirements for protecting underground pipelines from excavation-related damage; these requirements are located in 49 C.F.R. Part 196. Additionally, we excluded cases under the purview of the Headquarters.

Figure 8. Total Number of Cases the Pipeline and Hazardous Materials Safety Administration (PHMSA) Initiated byEnforcement Action Type, from 2010 to 2020.

Figure 8 displays the same data, but organized by enforcement action type per year for all regions. Warning Letters, Notices of Amendment, andNotices of Probable Violation remain the primary ways through which PHMSA enforces the pipeline safety laws and regulations.

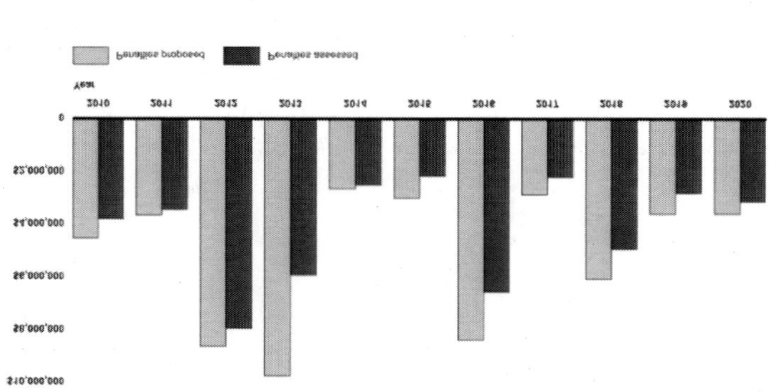

Source: GAO analysis of PHMSA enforcement data. | GAO-23-105537.

Notes: Civil monetary penalties may be proposed for probable violations identified during inspections, investigations, or other means. Proposed penalties are communicated to operators and the public in Notices of Probable Violation. An operator has the opportunity to contest the allegations, request a hearing, and submit written explanations that it believes may warrant mitigation or elimination of the proposed civil penalty before the penalty, if any, is assessed in the Final Order. PHMSA has discretion to determine the proposed and assessed amounts of a civil penalty. Thus, a proposed civil penalty may be greater than the assessed amount or not assessed at all. The factors that PHMSA considers in making these determinations are listed in its regulations. 49 C.F.R. § 190.225. PHMSA will not assess any penalties if it withdraws all of the allegations in a notice of probable violation prior to issuing a final order or if it determines that the operator did not violate a pipeline safety statute, regulation, or PHMSA order issued under them in its final order.

For this analysis, we excluded cases that involved liquefied natural gas (LNG) facilities and excavators that did not comply with PHMSA's requirements for protecting underground pipelines from excavation-related damage, which are located in 49 C.F.R. Part 196. Additionally, we excluded cases under the purview of the headquarters.

Figure 9. Total Amount of Penalties the Pipeline and Hazardous Materials Safety Administration (PHMSA) Proposed versus Assessed, from 2010 to 2020.

In addition to these data, we also analyzed the changes in the total amounts of civil monetary penalties proposed and assessed by PHMSA for Notice of Probable Violation cases.[26] Figure 9 displays the differences between these penalty amounts. On average, PHMSA pursued 41 Notice of Probable Violation cases with proposed civil monetary penalties a year from 2010 to 2020. Of these, an average of 13 cases a year saw a decrease in the amount of civil monetary penalties assessed versus the amount PHMSA proposed.

---

[26] Civil monetary penalties may be proposed for probable violations identified during inspections, investigations, or other means. Proposed penalties are communicated to operators and the public in Notices of Probable Violation. An operator has the opportunity to contest the allegations, request a hearing, and submit written explanations that it believes may warrant mitigation or elimination of the proposed civil penalty before the penalty, if any, is assessed in the Final Order. PHMSA has discretion to determine the proposed and assessed amounts of a civil penalty. Thus, a proposed civil penalty may be greater than the assessed amount, or not assessed at all. The factors that PHMSA considers in making these determinations are listed in its regulations. 49 C.F.R.§ 190.225. PHMSA will not assess any penalties if it withdraws all of the allegations in a notice of probable violation prior to issuing a final order, or if it determines that the operator did not violate a pipeline safety statute, regulation, or PHMSA order issued under them in its final order.

Appendix II: Comments from the Department of Transportation

U.S. Department of Transportation
Office of the Secretary of Transportation

Assistant Secretary for Administration

1200 New Jersey Avenue, SE
Washington, DC 20590

October 25, 2022

Ms. Elizabeth Repko
Director, Physical Infrastructure
U.S. Government Accountability Office (GAO)
441 G Street NW
Washington, DC 20548

Dear Ms. Repko:

The Pipeline and Hazardous Materials Safety Administration's (PHMSA) mission is to protect people and the environment by advancing the safe transportation of energy and other hazardous materials that are essential to our daily lives. To do this, the agency establishes national policy, sets and enforces standards, educates, and conducts research to prevent incidents. To provide greater public access to enforcement activities, PHMSA's Pipeline Enforcement web site makes available extensive enforcement records and information to the regulated community, the public, partner state agencies, and other stakeholders. PHMSA ensures accurate and complete enforcement records with centralized control using stringent data entry and quality control procedures.

Upon review of the draft report, PHMSA concurs with GAO's three recommendations to update PHMSA's Enforcement Transparency website to: (1) provide complete, machine-readable data that can be downloaded in bulk and in selected subsets; (2) be interoperable with other relevant datasets, such as PHMSA's accident/incident dataset (e.g., by adding a unique identifier variable); and (3) provide a search function that allows users to find relevant information, which could include the ability to search by the regulation violated. We will provide a detailed response to each recommendation within 180 days of the final report's issuance.

PHMSA appreciates the opportunity to respond to the GAO draft report. Please contact Gary Middleton, Director of Audit Relations and Program Improvement at (202) 366-6512 or gary.middleton@dot.gov with any questions or if GAO would like additional information.

Sincerely,

Philip McNamara
Assistant Secretary for Administration

Appendix III: GAO Contact and Staff Acknowledgments

GAO Contact

Elizabeth Repko, (202) 512-2834 or RepkoE@gao.gov.

Staff Acknowledgments

In addition to the contact above, Sara Vermillion (Assistant Director);Catrin Jones (Analyst-in-Charge); Sharon Dyer; Lauren Wice; Alicia Loucks; Mary-Catherine P. Overcash; Malika Rice; and Gina Hoovermade key contributions to this report.

GAO's Mission

The Government Accountability Office, the audit, evaluation, and investigative arm of Congress, exists to support Congress in meeting its constitutional responsibilities and to help improve the performance and accountability of the federal government for the American people. GAO examines the use of public funds; evaluates federal programs and policies; and provides analyses, recommendations, and other assistance to help Congress make informed oversight, policy, and funding decisions. GAO's commitment to good government is reflected in its core values of accountability, integrity, and reliability.

Obtaining Copies of GAO Reports and Testimony

The fastest and easiest way to obtain copies of GAO documents at no cost is through our website. Each weekday afternoon, GAO posts on its website newly released reports, testimony, and correspondence. You can also subscribe to GAO's email updates to receive notification of newly posted products.

Order by Phone

The price of each GAO publication reflects GAO's actual cost of production and distribution and depends on the number of pages in the publication and whether the publication is printed in color or black and white. Pricing and ordering information is posted on GAO's website, https://www.gao.gov/ordering.htm.

Place orders by calling (202) 512-6000, toll free (866) 801-7077, or

TDD (202) 512-2537.

Orders may be paid for using American Express, Discover Card, MasterCard, Visa, check, or money order. Call for additional information.

Connect with GAO

Connect with GAO on Facebook, Flickr, Twitter, and YouTube. Subscribe to our RSS Feeds or Email Updates. Listen to our Podcasts. Visit GAO on the web at https://www.gao.gov.

To Report Fraud, Waste, and Abuse in Federal Programs

Contact FraudNet:

Website: https://www.gao.gov/about/what-gao-does/fraudnet Automated answering system: (800) 424-5454 or (202) 512-7700

Congressional Relations

A. Nicole Clowers, Managing Director, ClowersA@gao.gov, (202) 512-4400, U.S. Government Accountability Office, 441 G Street NW, Room 7125, Washington, DC 20548

Public Affairs

Chuck Young, Managing Director, youngc1@gao.gov, (202) 512-4800

U.S. Government Accountability Office, 441 G Street NW, Room 7149 Washington, DC 20548

Strategic Planning and External Liaison

Stephen J. Sanford, Managing Director, spel@gao.gov, (202) 512-4707

U.S. Government Accountability Office, 441 G Street NW, Room 7814, Washington, DC 20548

Chapter 3

DOT's Federal Pipeline Safety Program: Background and Issues for Congress*

Paul W. Parfomak

Summary

The U.S. energy pipeline network includes approximately 3.3 million miles of onshore pipeline transporting natural gas, crude oil, and other hazardous liquids. Over the past decade, major safety incidents in California, Massachusetts, Mississippi, and other states have drawn criticism from stakeholders and have raised concerns in Congress about pipeline safety regulation. The 2021 ransomware attack on the Colonial Pipeline has also drawn attention to federal pipeline security activities, including agency roles and the linkage between pipeline safety and security.

The federal safety program for onshore pipelines is administered by the Department of Transportation's Pipeline and Hazardous Materials Safety Administration (PHMSA), which relies heavily on state partnerships for inspection and enforcement. PHMSA's pipeline safety program is authorized through FY2023 under the Protecting our Infrastructure of Pipelines and Enhancing Safety Act of 2020 (PIPES Act, P.L. 116-260, Div. R). President Biden's requested FY2024 budget for pipeline safety is $228.23 million, roughly 20% above the FY2023 budget authority.

The FY2024 request includes $89.56 million for grants to fund state pipeline inspection and damage prevention programs, up from $68.06 million in FY2023. The Infrastructure Investment and Jobs Act (IIJA, P.L. 117-58) appropriated $200 million annually through FY2026 for

* This is an edited, reformatted and augmented version of Congressional Research Service Publication No. R44201 Updated March 31, 2023.

In: Pipeline Safety: Background, Practices and Regulatory Changes
Editor: Michael E. Massaro
ISBN: 979-8-89530-058-9

PHMSA's new Natural Gas Distribution Infrastructure Safety and Modernization Grant Program.

To promote regulatory compliance, PHMSA conducts programmatic inspections of management systems and procedures; inspects facilities and construction; investigates safety incidents; and maintains a dialogue with pipeline operators. The agency clarifies its expectations through orders, guidance manuals, and public meetings. It also administers a pipeline safety research and development (R&D) program to address emerging risks and new technologies. PHMSA works with the Transportation Security Administration (TSA) on pipeline security and incident response.

As oversight of PHMSA's pipeline safety program continues, Congress may examine PHMSA staffing, which faces persistent shortfalls affecting the agency's ability to inspect pipelines and revise its regulations. Other potential topics for congressional oversight could include

- the implementation of PHMSA's new distribution modernization grant program;
- the effects of the agency's 2021 rule for natural gas gathering lines, bringing 425,000 miles of gathering lines under regulation;
- PHMSA's implementation of the PIPES Act mandate expanding PHMSA's traditional safety mission to include climate considerations;
- what role PHMSA might play in any future TSA pipeline security initiatives;
- updates to outdated safety standards for liquefied natural gas facilities, and pipelines carrying carbon dioxide, hydrogen, or hydrogen-methane blends;
- PHMSA's issuance and oversight of standards exemptions via Special Permits; and
- PHMSA's implementation and coordination of pipeline safety R&D through its own grants, operator demonstrations, and programs at other federal agencies.

In addition to these issues, Congress may assess how the many elements of U.S. pipeline safety fit together in the nation's overall approach to protecting the public and the environment. Pipeline safety necessarily involves various groups: federal and state agencies, tribal governments, pipeline associations, large and small pipeline operators, local communities, and other interest groups. Reviewing how these groups work together to achieve common goals or resolve conflicting approaches could be an overarching concern for Congress.

Introduction

The U.S. energy pipeline network is integral to the nation's energy supply and provides vital links to other critical infrastructure, such as power plants, airports, and military bases. These pipelines are geographically widespread, running alternately through remote and densely populated regions—from Arctic Alaska to the Gulf of Mexico and nearly everywhere in between (Figure 1). Because energy pipelines carry volatile, flammable, or toxic materials, they have the potential to injure the public, destroy property, and harm the environment. Although they are considered an efficient and comparatively safe means of transport, pipeline systems are also vulnerable to accidents, operational failure, and malicious attacks. Recent major incidents in California, Massachusetts, and Mississippi, among other places, have demonstrated the risks of pipeline failure and have heightened congressional concern about U.S. pipeline safety. A 2021 cyberattack on the Colonial Pipeline likewise demonstrated the economic impacts of a major pipeline disruption and put a focus on the linkage between pipeline safety and security.

The federal safety program for onshore pipelines resides primarily within the Department of Transportation's (DOT's) Pipeline and Hazardous Materials Safety Administration (PHMSA), although its inspection and enforcement activities rely heavily upon partnerships with the states. Together, the federal and state pipeline safety agencies administer a comprehensive set of regulatory authorities that continues to evolve. DOT's pipeline safety program is authorized through the fiscal year ending September 30, 2023, under the Protecting our Infrastructure of Pipelines and Enhancing Safety Act of 2020 (PIPES Act, P.L. 116-260, Div. R) signed by President Trump on December 27, 2020.

This report reviews the history and role of the federal program for pipeline safety, including a discussion of pipeline safety trends and major accidents. It discusses significant regulatory changes in reauthorization statutes and summarizes ongoing developments in key policy areas. It discusses PHMSA's relationship with other federal agencies involved in pipeline safety. Although pipeline security is not mainly under PHMSA's jurisdiction, the report examines the agency's role in pipeline security and its recent work on security-related issues with other agencies.

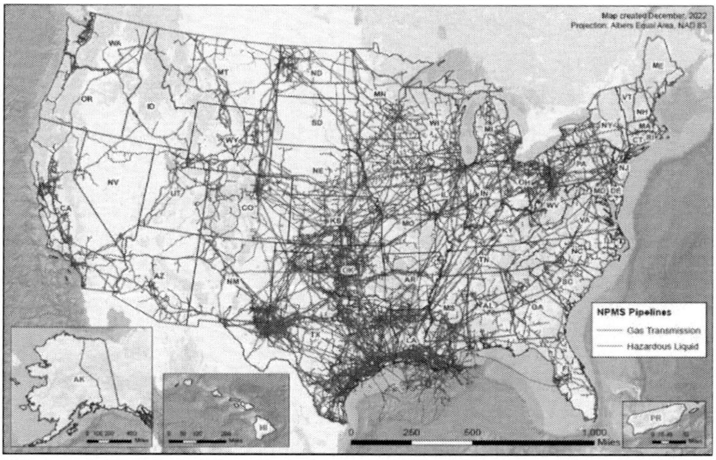

Source: National Pipeline Mapping System (NPMS), December 19, 2022, https:// www.npms.phmsa.dot.gov/ Documents/NPMS_Pipelines_Map.pdf.

Notes: Map does not show gas distribution or gas gathering pipelines. Hazardous liquids primarily include crude oil, gasoline, jet fuel, diesel fuel, home heating oil, propane, and butane. Other hazardous liquids transported by pipeline include anhydrous ammonia, carbon dioxide, kerosene, liquefied ethylene, and petrochemical feedstock.

Figure 1. U.S. Natural Gas Transmission and Hazardous Liquid Pipelines.

The U.S. Pipeline Network

The onshore U.S. energy pipeline network is composed of approximately 3.3 million miles of pipeline transporting natural gas, oil, and other hazardous liquids (Table 1). Of the nation's approximately half-million miles of long-distance transmission pipeline, roughly 230,000 miles carry hazardous liquids—over 80% of the nation's crude oil and refined products—along with other products.[1] It also contains some 47,000 miles of crude oil gathering pipelines, which connect extraction wells to processing facilities prior to long-distance shipment. The U.S. natural gas pipeline network consists of around 302,000 miles of transmission and 434,000 miles of gathering lines. The natural gas transmission pipelines feed around 2.3 million miles of regional

[1] Bureau of Transportation Statistics, "Crude Oil and Petroleum Products Transported in the United States by Mode," https://www.bts.gov/content/crude-oil-and-petroleum-products-transported-united-states-mode, accessed January 10, 2022.

pipeline mains in some 1,500 local distribution networks serving over 70 million customers.[2]

Table 1. U.S. hazardous liquid and natural gas pipeline mileage 2021

Category	Miles
Hazardous Liquids Transmission	229,958
Hazardous Liquids Gathering	47,126
Natural Gas Transmission	301,502
Natural Gas Gathering	434,076
Natural Gas Distribution Mains and Service Lines	2,301,090
TOTAL	3,313,752

Sources: Hazardous liquids transmission, natural gas transmission, and natural gas distribution mains and service lines mileage is from PHMSA, "Annual Report Mileage Summary Statistics," web tables, February 6, 2023. http://www.phmsa.dot.gov/portal/site/PHMSA/menuitem.7c371785a639f2e55cf2031050248a0c/?vgnextoid=3b6c03347e4d8210VgnVCM1000001ecb7898RCRD&vgnextchannel=3b6c03347e4d8210VgnVCM1000001ecb7898RCRD&vgnextfmt=print; Hazardous liquids and natural gas gathering lines mileage is from Environmental Protection Agency, "Inventory of U.S. Greenhouse Gas Emissions and Sinks 1990-2020: Updates Under Consideration for Activity Data," memorandum, September 2021, p. 3, https://www.epa.gov/system/files/documents/2021-09/2022-ghgi-update-activity-data_sept-2021.pdf.

Notes: Hazardous liquids primarily include crude oil, gasoline, jet fuel, diesel fuel, home heating oil, propane, and butane. Other hazardous liquids transported by pipeline include anhydrous ammonia, carbon dioxide, kerosene, liquefied ethylene, and petrochemical feedstock. Hazardous liquids gathering mileage is for crude oil pipelines.

Natural gas pipelines also connect to 168 active liquefied natural gas (LNG) storage sites, as well as underground storage facilities, both of which can augment pipeline gas supplies during peak demand periods.[3]

---

[2] PHMSA, "Annual Report Mileage for Gas Distribution Systems," January 4, 2022, https://www.phmsa.dot.gov/data-and-statistics/pipeline/annual-report-mileage-gas-distribution-systems.

[3] PHMSA, "Liquefied Natural Gas (LNG) Facilities and Total Storage Capacities," February 6, 2023, https://www.phmsa.dot.gov/data-and-statistics/pipeline/liquefied-natural-gas-lng-facilities-and-total-storage-capacities.

Safety in the Pipeline Industry

Uncontrolled pipeline releases can result from a variety of causes, including third-party excavation, corrosion, mechanical failure, control system failure, operator error, and malicious acts. Natural forces, such as floods and earthquakes, can also damage pipelines. Taken as a whole, releases from pipelines cause few annual injuries or fatalities compared to other product transportation modes.[4] According to PHMSA statistics, there were, on average, 12 deaths and 58 injuries caused by 29 pipeline incidents annually in all U.S. pipeline systems from 2010 through 2021.[5] After a steady decline in the incidents causing injuries or fatalities between 2010 and 2013, the average incident count has fluctuated, although it remained relatively flat between 2019 and 2021 (Figure 2). A total of 26 serious pipeline incidents were reported for 2021.

(Annual "Serious" Incidents)

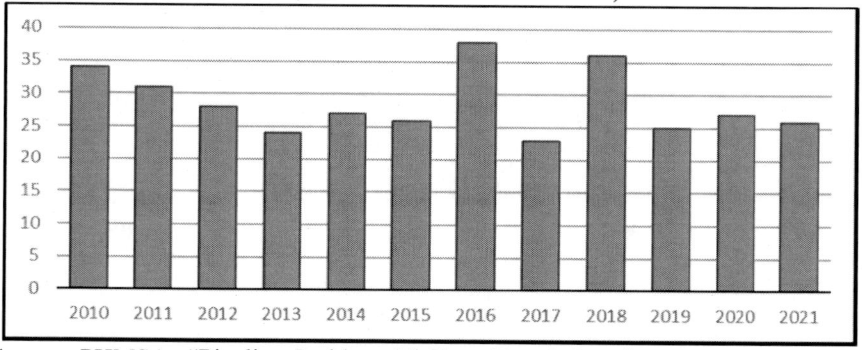

Source: PHMSA, "Pipeline Incident 20 Year Trends," online database, accessed March 7, 2023, https://www.phmsa.dot.gov/data-and-statistics/pipeline/pipeline-incident-20-year-trends.

Note: PHMSA defines "serious" incidents as those causing a fatality or injury requiring inpatient hospitalization.

Figure 2. Pipeline Incidents Causing Injuries or Fatalities 2010-2021.

Apart from injury to people, some accidents may cause local environmental damage or other physical impacts, which may be significant,

[4] Bureau of Transportation Statistics, *National Transportation Statistics: 2021*, Table 2-4.

[5] PHMSA, PHMSA, "Pipeline Incident 20 Year Trends," accessed March 7, 2023, https://www.phmsa.dot.gov/data- and-statistics/pipeline/pipeline-incident-20-year-trends.

particularly in the case of oil spills or fires. PHMSA requires the reporting of such incidents involving

- $50,000 or more in total costs, measured in 1984 dollars,
- highly volatile liquid releases of 5 barrels or more or other liquid releases of 50 barrels or more, or
- liquid releases resulting in an unintentional fire or explosion.[6]

On average there were 264 such "significant" incidents (those not involving injury or fatality) per year from 2010 through 2021. The average significant incident count has fluctuated since 2010, with no clear overall trend, although incidents declined from 2019 to 2021 (Figure 3). A total of 241 significant pipeline incidents were reported for 2021. It should be noted that federally regulated pipeline mileage overall rose approximately 9% over this period; neither the annual statistics for injury nor environmental incidents are adjusted on a per-mile basis.[7]

Although pipeline releases have caused relatively few fatalities in absolute numbers, a single pipeline accident can be catastrophic in terms of public safety and environmental damage. For example, in 2015, the Aliso Canyon Underground Storage Facility near the Porter Ranch community in Los Angeles County, CA, began experiencing an uncontrolled natural gas leak that ultimately released an estimated 109,000 metric tons of methane, a potent greenhouse gas (GHG).[8] The risk to safety from the fugitive methane and the presence of odorants and other chemicals in the gas led to the temporary relocation of over 8,000 households and two schools in nearby Porter Ranch. In 2018, overpressure in a natural gas distribution main in Merrimack Valley, MA, killed one person, injured 21 others, damaged 131 structures, and caused 30,000 residents to evacuate their homes for several days.[9] Such incidents have generated persistent scrutiny of pipeline risks and have increased federal, state, and community activity related to pipeline safety.

---

[6] PHMSA, "Pipeline Incident Flagged Files," February 28, 2023, https://www.phmsa.dot.gov/data-and-statistics/pipeline/pipeline-incident-flagged-files. The definition excludes natural gas distribution incidents caused by a nearby fire or explosion impacting the pipeline system.

[7] For detailed annual pipeline mileage statistics, see PHMSA, "Annual Report Mileage Summary Statistics," March 6, 2023, https://www.phmsa.dot.gov/data-and-statistics/pipeline/annual-report-mileage-summary-statistics.

[8] County of Los Angeles, Department of Public Health, "Aliso Canyon Disaster Health Research Study," 2021, http://publichealth.lacounty.gov/eh/healthresearch/back ground.htm.

[9] NTSB, "Pipeline Over-Pressure of a Columbia Gas of Massachusetts Low-Pressure Natural Gas Distribution System Merrimack Valley, Massachusetts, September 13, 2018," preliminary report, PLD18MR003, October 10, 2018.

(Annual "Significant" Incidents)

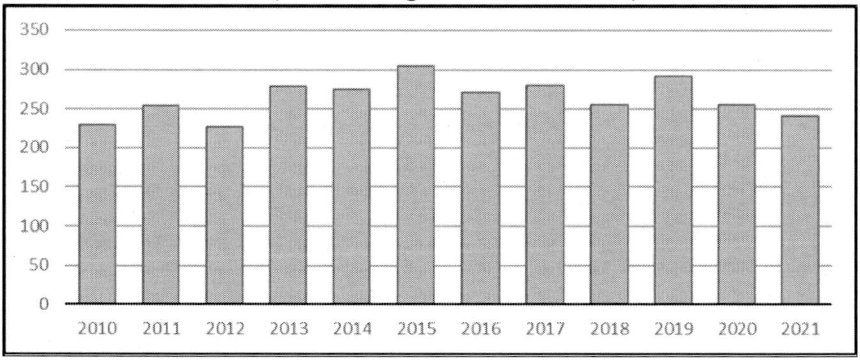

Source: PHMSA, "Pipeline Incident 20 Year Trends," online database, accessed March 7, 2023, https://www.phmsa.dot.gov/data-and-statistics/pipeline/pipeline-incident-20-year-trends.

Notes: Includes "significant" incidents, with $50,000 or more in total costs (1984 dollars), highly volatile liquid releases of 5 barrels or more or other liquid releases of 50 barrels or more, or liquid releases resulting in an unintentional fire or explosion. Excludes incidents causing a fatality or injury requiring inpatient hospitalization.

Figure 3. Pipeline Incidents Causing Environmental or Property Damage 2010-2021.

Notable Pipeline Safety Incidents since 2010

- *2010*—A pipeline spill in Marshall, MI, released 19,500 barrels of crude oil into a Kalamazoo River tributary.
- *2010*—A pipeline explosion in San Bruno, CA, killed 8 people, injured 60 others, and destroyed 37 homes.
- *2011*—An explosion caused by a natural gas pipeline in Allentown, PA, killed 5 people, damaged 50 buildings, and caused 500 people to be evacuated.
- *2011*—A pipeline near Laurel, MT, spilled an estimated 1,000 barrels of crude oil into the Yellowstone River.
- *2012*—A natural gas pipeline explosion in Springfield, MA, injured 21 people and damaged over 12 buildings.
- *2014*—An explosion caused by a natural gas distribution pipeline in New York City killed 8 people, injured 50 others, and destroyed two 5-story buildings.

- *2015*—A pipeline in Santa Barbara County, CA, spilled 3,400 barrels of crude oil, including 500 barrels reaching Refugio State Beach on the Pacific Ocean.
- *2015*—The Aliso Canyon natural gas storage facility in Los Angeles County, CA, released 5.4 billion cubic feet of gas, causing the temporary relocation of over 2,000 households and two schools in Porter Ranch.
- *2016*—An explosion caused by a natural gas distribution pipeline in Canton, OH, killed one person, injured 11 others, and damaged over 50 buildings.
- *2018*—Explosions and fires caused by natural gas distribution pipelines in Merrimack Valley, MA, killed one person, injured 21 others, damaged 131 structures, and caused 30,000 residents to evacuate.
- *2020*—A carbon dioxide pipeline ruptured near Satartia, MS, leading to a local evacuation and causing 45 people to be hospitalized.
- *2021*—An underwater oil pipeline off of Long Beach, CA, damaged by a ship's anchor spilled over 500 barrels of oil into San Pedro Bay.
- *2022*—An explosion and fire at an LNG export terminal in Freeport, TX, resulted in a months-long facility shutdown and temporarily stopped approximately 20% of U.S. LNG exports.
- *2022*—A pipeline rupture near Washington, KS, spilled an estimated 13,000 barrels of crude oil, some of which reached a nearby creek.
- *2023*—A natural gas pipeline-related explosion and fire at a West Reading, PA, factory killed 7 people, caused 10 others to be hospitalized, and damaged several other nearby buildings.

PHMSA's Pipeline Safety Program

PHMSA has the primary responsibility for the formulation, administration, and oversight of onshore pipeline safety regulations in the United States. The agency does so through its Office of Pipeline Safety (OPS), whose functions include oversight of pipeline operators, support of state pipeline safety agencies, and cooperation with other federal agencies that have pipeline safety responsibilities. The latter include the Department of Interior's Bureau of Safety and Environmental Enforcement (BSEE), which regulates offshore oil and natural gas facilities, and the Federal Energy Regulatory Commission (FERC), which has siting authority for interstate natural gas pipelines. PHMSA also cooperates with the National Transportation Safety Board (NTSB), an independent agency that investigates accidents and issues safety recommendations.

Pipeline and Hazardous Materials Safety Administration

The Natural Gas Pipeline Safety Act of 1968 (P.L. 90-481) and the Hazardous Liquid Pipeline Act of 1979 (P.L. 96-129) are the principal acts establishing the federal role in pipeline safety. Under both statutes, the Secretary of Transportation has primary authority to regulate key aspects of pipeline safety: design, construction, operation and maintenance, and spill response planning. Pipeline safety regulations are covered in Title 49 of the Code of Federal Regulations.[10]

Organization and Funding

As of February 12, 2023, PHMSA's organizational chart listed 285 full-time equivalent (FTE) staff in OPS—including 153 pipeline safety inspectors.[11] There are also 33 agency positions outside of OPS that support certain pipeline safety functions.[12] In addition to federal staff, PHMSA's enabling legislation allows the agency to delegate authority to intrastate pipeline safety offices, enabling them to act as "agents" administering interstate pipeline safety programs (excluding enforcement) for those sections of interstate pipelines within their boundaries.[13] According to the DOT, "PHMSA relies on state inspectors for inspecting the vast network of intrastate pipelines."[14] A few states serve as agents for inspection of interstate pipelines as well. There are 436 state inspectors in 2023.[15] PHMSA may reimburse states for up to 80% of their pipeline safety expenditures through State Pipeline Safety Grants. In 2020 (the latest year with published data) actual grant awards to states covered approximately 70% of state expenditures, on average.[16] PHMSA may also

---

[10] Safety and security of LNG facilities used in gas pipeline transportation is regulated under Title 49, Part 193, of the *Code of Federal Regulations.*

[11] PHMSA, "Office of Pipeline Safety Organization Chart," web page, February 27, 2023, https://www.phmsa.dot.gov/ about-phmsa/offices/office-pipeline-safety-organization-chart.

[12] Damon Hill, PHMSA, personal communication, March 23, 2023. Those staff include attorneys, data analysts, information technology specialists, and regulatory specialists required for certain enforcement actions, promulgating regulations, issuing pipeline safety grants, and issuing agreements for pipeline safety research and development.

[13] 49 U.S.C. §60107.

[14] DOT, *Budget Estimates Fiscal Year 2020, Pipeline and Hazardous Materials Safety Administration*, 2019, p. 24, https://www.transportation.gov/sites/dot.gov/files/docs/mission/budget/334301/fy-2020-phmsa-budget-508- compliant.pdf.

[15] DOT, *Budget Estimates Fiscal Year 2024, Pipeline and Hazardous Materials Safety Administration*, 2023, p. 35, https://www.transportation.gov/sites/dot.gov/files/2023-03/ PHMSA_FY_2024_President_Budget_508.pdf.

[16] PHMSA, "Base Grant Payment Info 2008-2020," February 3, 2021, https://www. phmsa.dot.gov/grants/pipeline/ base-grant-payment-info-2008-2020.

fund states through Underground Natural Storage Grants, State Damage Prevention Grants, State One-Call Grants, and Natural Gas Distribution Infrastructure Safety and Modernization Grants, further discussed below.

(Millions of Dollars)

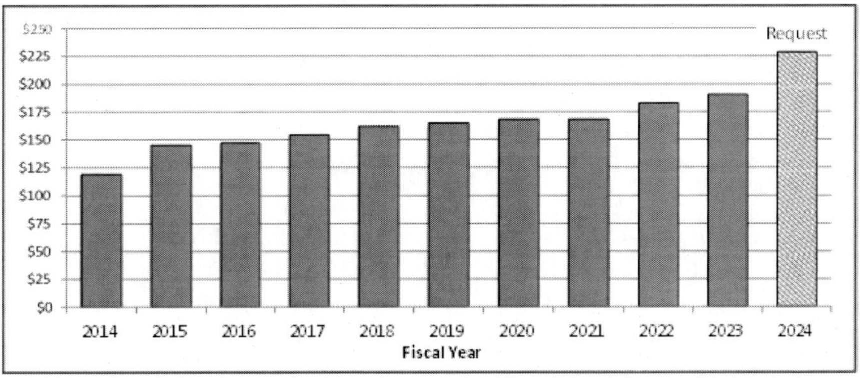

Source: U.S. Office of Management and Budget, Budget of the United States Government, Appendix, Fiscal Years 2014 through 2024, "Pipeline Safety," Line 1900 "Budget authority (total)."

Notes: Column values are "actual" budget authority totals except for 2023, which is "enacted," and 2024, which is reported in the Biden Administration's FY2024 budget appendix. Values are not adjusted for inflation.

Figure 4. PHMSA Pipeline Safety Total Annual Budget Authority 2014-2024.

PHMSA's pipeline safety program is funded primarily by user fees assessed on a per-mile basis on each regulated pipeline operator.[17] The agency's total annual budget authority has grown over the last decade (*Figure 4*). The Biden Administration's requested pipeline safety budget for PHMSA in FY2024 is $228.23 million, roughly 20% greater than the FY2023 budget authority. The FY2024 request includes $89.56 million for grant programs funding state pipeline inspections and damage prevention, up from $68.06 million in FY2023.

Regulatory Activities

PHMSA uses a variety of strategies to promote compliance with its safety standards. The agency conducts programmatic inspections of management systems, procedures, and processes; conducts physical inspections of facilities and construction projects; investigates safety incidents; and maintains a

[17] 49 U.S.C. §60125.

dialogue with pipeline operators. The agency clarifies its regulatory expectations through published protocols and regulatory orders, guidance manuals, and public meetings.

In 1997, PHMSA began requiring industry to implement "integrity management" programs on pipeline segments near "high consequence areas." Integrity management provides for continual evaluation of pipeline condition, assessment of risks to the pipeline, inspection or testing, data analysis, and follow-up repair as well as preventive or mitigative actions. High consequence areas include population centers, commercially navigable waters, and environmentally sensitive areas, such as drinking water supplies or ecological reserves. The integrity management approach is intended to prioritize resources to locations of highest consequence rather than applying uniform treatment to the entire pipeline network. PHMSA made integrity management programs mandatory for most oil pipeline operators with 500 or more miles of regulated pipeline as of March 31, 2001 (49 C.F.R. §195). Congress subsequently mandated the expansion of integrity management to natural gas pipelines and has continued to make other significant changes to federal pipeline safety requirements through PHMSA budget reauthorizations as discussed below.

Regulation of Offshore Pipelines

Offshore pipelines are regulated primarily by BSEE, which is responsible for the safety and environmental oversight of oil and gas operations as well as oil spill response on the Outer Continental Shelf.[18] PHMSA shares with BSEE oversight of certain offshore pipeline facilities. Under the terms of a December 2020 Memorandum of Understanding (MOU) between the two agencies, PHMSA is responsible for "all OCS pipelines beginning downstream of the point at which operating responsibility transfers from a producing operator to a transporting operator, or downstream of the last valve on the last production facility on the OCS for pipelines that cross into State waters."[19] In addition,

[18] BSEE was established in 2011 under a secretarial order reorganizing the former Minerals Management Service. See Secretary of the Interior, Order No. 3299, Amendment No. 2, August 29, 2011, https://www.doi.gov/sites/doi.gov/files/ elips/documents/3299a2-establishment_of_the_bureau_of_ocean_energy_management_the_bureau_of_safety_and_e nvironmental_enforcement_and_the_office_of_natural_resources_revenue.pdf. BSEE's regulations are found under Title 30 (Mineral Resources) of the *Code of Federal Regulations*.

[19] BSEE and PHMSA, "Memorandum of Understanding Between the U.S. Department of Transportation and the U.S. Department of the Interior Regarding Outer Continental Shelf Pipelines," December 22, 2020, p. 3, https://www.bsee.gov/sites/bsee.gov/files/mou-est-17430-doi-dot-outer-continental-shelf-pipelines-mou-2020-12- 22.pdf.

BSEE regulations allow a producer to petition to have its pipeline operate under PHMSA regulations for pipeline design, construction, operation, and maintenance.[20] Likewise, a transporter who operates a PHMSA-regulated pipeline may petition to operate under BSEE regulations for pipeline operation and maintenance.[21] Policy issues related primarily to BSEE or to pipelines under its jurisdiction are outside the scope of this report.

Pipeline Safety Enforcement

PHMSA relies upon a range of enforcement actions, including administrative actions such as safety orders and civil penalties, to try to ensure that operators correct safety violations and take measures to preclude future safety problems. From 2018 through 2022, PHMSA initiated 1,108 enforcement actions against pipeline operators.[22] Of these cases, 352 resulted in notices of probable violation, which allege specific regulatory violations, and 16 resulted in corrective action orders, which "usually address urgent situations arising out of an accident, spill, or other significant, immediate, or imminent safety or environmental concern."[23] Civil penalties proposed by PHMSA for safety violations during this period totaled approximately $37 million.[24] PHMSA also conducts accident investigations and system-wide reviews focusing on high-risk operational or procedural problems and areas of the pipeline near sensitive environmental areas, high-density populations, or navigable waters.

Reauthorization and Pipeline Safety Statutes

The PIPES Act was preceded by a periodic series of pipeline safety statutes, each of which reauthorized funding for PHMSA's pipeline safety program and included other provisions related to PHMSA's authorities, administration, or regulatory activities.

[20] 30 C.F.R §250.1000(c)(12).

[21] 30 C.F.R §250.1000(c)(13).

[22] PHMSA, "PHMSA Pipeline Safety Program: Summary of Enforcement Actions," March 6, 2023, http://primis.phmsa.dot.gov/comm/reports/enforce/Actions_opid_0.html?Nocache =8828.

[23] PHMSA, "PHMSA Pipeline Safety Program: Summary of Enforcement Actions."

[24] PHMSA, "PHMSA Pipeline Safety Program: Summary of Cases Involving Civil Penalties," March 6, 2023, http://primis.phmsa.dot.gov/comm/reports/enforce/CivilPenalty_opid_0.html?nocache=9288#_TP_1_tab_1. Proposed penalties may change in the resolution of a case.

Pipeline Safety Improvement Act of 2002

On December 12, 2002, President George W. Bush signed into law the Pipeline Safety Improvement Act of 2002 (P.L. 107-355). The act strengthened federal pipeline safety programs, state oversight of pipeline operators, and public education regarding pipeline safety.[25] Among other provisions, P.L. 107-355 required operators of regulated natural gas pipelines in high consequence areas to conduct risk analysis and implement integrity management programs similar to those required for oil pipelines. The act authorized DOT to order safety actions for pipelines with potential safety problems and increased violation penalties. The act streamlined the permitting process for emergency pipeline restoration by establishing an interagency committee—including the DOT, the Environmental Protection Agency, the Bureau of Land Management, FERC, and other agencies—to ensure coordinated review and permitting of pipeline repairs. The act required DOT to study ways to limit pipeline safety risks from population encroachment and ways to preserve environmental resources in pipeline rights-of-way.

P.L. 107-355 also included provisions for public education, grants for community pipeline safety studies, "whistleblower" and other employee protection, employee qualification programs, and mapping data submission.

Pipeline Inspection, Protection, Enforcement, and Safety Act of 2006

On December 29, 2006, President Bush signed into law the Pipeline Inspection, Protection, Enforcement and Safety Act of 2006 (P.L. 109-468). The main provisions of the act address pipeline damage prevention, integrity management, corrosion control, and enforcement transparency. The act created a national focus on pipeline damage prevention through grants to states for improving damage prevention programs, establishing 811 as the national "call before you dig" one-call telephone number, and giving PHMSA limited "backstop" authority to conduct civil enforcement against one-call violators in states that have failed to conduct such enforcement. The act mandated the promulgation by PHMSA of minimum standards for integrity management programs for natural gas distribution pipelines.[26] It also mandated a review of

[25] P.L. 107-355 encourages the implementation of state "one-call" excavation notification programs (§2) and allows states to enforce "one-call" program requirements. The act expands criminal responsibility for pipeline damage to cases where damage was not caused "knowingly and willfully" (§3). The act adds provisions for ending federal-state pipeline oversight partnerships if states do not comply with federal requirements (§4).

[26] PHMSA issued final regulations requiring operators of natural gas distribution pipelines to adopt integrity management programs similar to existing requirements for gas transmission pipelines on December 4, 2009.

the adequacy of federal pipeline safety regulations related to internal corrosion control and required PHMSA to increase the transparency of enforcement actions by issuing monthly summaries including violation and penalty information and a mechanism for pipeline operators to make response information available to the public.

Pipeline Safety, Regulatory Certainty, and Job Creation Act of 2011

On January 3, 2012, President Obama signed the Pipeline Safety, Regulatory Certainty, and Job Creation Act of 2011 (Pipeline Safety Act, P.L. 112-90). The act contains a broad range of provisions addressing pipeline safety. Among the most significant are provisions to increase the number of federal pipeline safety inspectors, require automatic shutoff valves for transmission pipelines, mandate verification of maximum allowable operating pressure for gas transmission pipelines, increase civil penalties for pipeline safety violations, and mandate reviews of diluted bitumen pipeline regulation. Altogether, the act imposed 42 mandates on PHMSA regarding studies, rules, maps, and other elements of the federal pipeline safety program. P.L. 112-90 authorized the federal pipeline safety program through the fiscal year ending September 30, 2015.

Protecting Our Infrastructure of Pipelines and Enhancing Safety Act of 2016

On June 22, 2016, President Obama signed the Protecting our Infrastructure of Pipelines and Enhancing Safety Act of 2016 (P.L. 114-183). Among other provisions, the act requires PHMSA to promulgate federal safety standards for underground natural gas storage facilities and grants PHMSA emergency order authority to address urgent "industry-wide safety conditions" without prior notice. The act also requires PHMSA to report regularly on the progress of outstanding statutory mandates. The act authorized the federal pipeline safety program through FY2019.

Protecting Our Infrastructure of Pipelines and Enhancing Safety Act of 2020

On December 27, 2020, President Trump signed the PIPES Act (P.L. 116-260, Div. R).[27] The act authorizes the federal pipeline safety program through FY2023. Among its key provisions, the act requires PHMSA to review and update its safety standards for large-scale LNG facilities, adopting a risk-based

[27] P.L. 116-260 is the Consolidated Appropriations Act, 2021.

regulatory approach. The act also imposes stricter standards for natural gas pipeline leak detection and repair, requiring repair of all leaks hazardous to human safety or the environment or with the potential to become hazardous. It also mandates new safety requirements for natural gas distribution systems in response to the 2018 Merrimack Valley incident.[28] These requirements include updates to distribution integrity management, emergency response plans to address over-pressurization risks, and a requirement for PHMSA to report on industry adoption of pipeline safety management systems. The act also includes provisions intended to help PHMSA attract and maintain a sufficient workforce of pipeline inspection and enforcement personnel.

In addition to the authorization in the PIPES Act, IIJA authorizes annual funding through FY2026 for a new Natural Gas Distribution Infrastructure Safety and Modernization Grant Program administered by PHMSA.[29]

Cooperation with FERC

One area related to pipeline safety not under PHMSA's primary jurisdiction is the siting approval of interstate natural gas pipelines, which is the responsibility of FERC. Companies building interstate natural gas pipelines must first obtain from FERC certificates of public convenience and necessity. (FERC does not oversee oil pipeline siting or construction.) FERC must also approve the abandonment of gas facility use and services. These approvals may include safety provisions with respect to pipeline routing, safety standards, and other factors.[30] In particular, pipeline and aboveground facilities associated with a proposed pipeline project must be designed in accordance with PHMSA's safety standards regarding material selection and qualification, design requirements, and protection from corrosion.[31]

PHMSA and FERC cooperate on pipeline safety-related matters according to an MOU signed in 1993. According to the MOU, PHMSA agrees to

- promptly alert FERC when safety activities may impact commission responsibilities,

---

[28] These provisions are included as the "Leonel Rondon Pipeline Safety Act," Title II of the PIPES Act.

[29] P.L. 117-58, Division J, Title VIII.

[30] In making permitting decisions for cross-border oil and natural gas pipelines, the State Department or FERC, respectively, must also consult with the Secretary of Transportation regarding pipeline safety, among other matters, in accordance with directives in Executive Order 13337.

[31] 18 C.F.R. §157.

- notify FERC of major accidents or significant enforcement actions involving pipelines under FERC's jurisdiction,
- refer to FERC any complaints and inquiries by state and local governments and the public about environmental or certificate matters related to FERC- jurisdictional pipelines, and
- when requested by FERC, review draft mitigation conditions considered by the commission for potential conflicts with PHMSA's regulations.

Under the MOU, FERC agrees to

- promptly alert PHMSA when the commission learns of an existing or potential safety problem involving natural gas transmission facilities;
- notify PHMSA of future pipeline construction;
- periodically provide PHMSA with updates to the environmental compliance inspection schedule and coordinate site inspections, upon request, with PHMSA officials;
- notify PHMSA when significant safety issues have been raised during the preparation of environmental assessments or environmental impact statements for pipeline projects; and
- refer to PHMSA complaints and inquiries made by state and local governments and the public involving safety matters related to FERC-jurisdictional pipelines.[32]

FERC may also serve as a member of PHMSA's Technical Pipeline Safety Standards Committee, which determines whether proposed safety regulations are technically feasible, reasonable, cost- effective, and practicable.

In April 2015, FERC issued a policy statement to provide "greater certainty regarding the ability of interstate natural gas pipelines to recover the costs of modernizing their facilities and infrastructure to enhance the efficient and safe operation of their systems."[33] FERC's policy statement was motivated by the commission's expectation that governmental safety and

[32] DOT and FERC, "Memorandum of Understanding Between the Department of Transportation and Federal Energy Regulatory Commission Regarding Natural Gas Transportation Facilities," January 15, 1993. Note that the MOU refers to DOT's Research and Special Programs Administration, the predecessor agency to PHMSA.

[33] FERC, *Cost Recovery Mechanisms for Modernization of Natural Gas Facilities*, 151 FERC ¶ 61,047, April 16, 2015, http://www.ferc.gov/whats-new/comm-meet/2015/041615/G-1.pdf.

environmental initiatives could cause greater safety and reliability costs for interstate gas pipeline systems.[34]

PHMSA and the NTSB

The NTSB is an independent federal agency charged with determining the probable cause of transportation incidents—including pipeline releases—and promoting transportation safety. The board's experts investigate significant incidents, develop factual records, and issue safety recommendations to prevent similar events from reoccurring. The NTSB has no statutory authority to regulate transportation, however, and it does not perform cost-benefit analyses of regulatory changes; its safety recommendations to industry or government agencies are not mandatory. Nonetheless, because of the board's strong reputation for thoroughness and objectivity, 82% of the NTSB's safety recommendations have been implemented across all transportation modes.[35]

In the pipeline sector, the NTSB's past safety recommendations have led to changes in pipeline safety regulation regarding one-call systems before excavation ("call before you dig"), use of pipeline internal inspection devices, facility response plan effectiveness, hydrostatic pressure testing of older pipelines, and other safety improvements.[36] As of March 10, 2023, the NTSB listed 15 open pipeline safety recommendations to PHMSA dating back to 2011. In all cases but one, the NTSB has classified these recommendations as "Open—Acceptable Response" or "Open—Acceptable Alternate Response" because they are being incorporated satisfactorily in ongoing PHMSA rulemakings or because PHMSA is implementing other measures to meet the same objectives. One recommendation is classified as "Open—Unacceptable Response," because the NTSB is not satisfied with PHMSA's actions to implement it. Detailed discussion of NTSB pipeline accident investigations and safety recommendations are publicly available through the NTSB's Case Analysis and Reporting Online online query tool.[37] In addition to making specific safety recommendations, the NTSB also comments on proposed changes to PHMSA's pipeline safety regulations, such as those involving pipeline hazard class locations and standards for valve installation and rupture detection, among other standards.[38]

---

[34] FERC, April 16, 2015, p. 1.

[35] NTSB, *Annual Report to Congress 2021*, 2022, p. 8. The 82% applies to recommendations closed by NTSB.

[36] NTSB, *Annual Report to Congress 2017*, 2018, p. 15.

[37] NTSB, "CAROL Query," online database, at https://data.ntsb.gov/carol-main-public/landing-page.

[38] NTSB, 2021, 41.

PHMSA's Role in Pipeline Security

Pipeline safety and security are distinct issues involving different threats, statutory authorities, and regulatory frameworks. Nonetheless, aspects of pipeline safety and security can be intertwined. PHMSA has historically played a significant role in pipeline security and continues to be involved in pipeline security oversight and incident response. The 2021 ransomware attack on the Colonial Pipeline Company, which disrupted gasoline supplies throughout the East Coast, elevated concern in Congress about federal oversight of pipeline security, including PHMSA's role within the nation's pipeline security framework.[39]

DOT's Early Role in Pipeline Security

DOT played the leading role in pipeline security through the late 1990s. Presidential Decision Directive 63 (PDD-63), issued during the Clinton Administration, assigned lead responsibility for pipeline security to DOT.[40] These responsibilities fell to OPS, at that time a part of DOT's Research and Special Programs Administration, because the agency was already addressing some elements of pipeline security in its role as safety regulator.[41] The DOT's pipeline (and LNG) safety regulations already included provisions related to physical security, such as requirements to protect surface facilities (e.g., pumping stations) from vandalism and unauthorized entry.[42] Other regulations required continuing surveillance, patrolling pipeline rights-of-way, damage prevention, and emergency procedures.[43]

On September 5, 2002, OPS circulated formal guidance developed in cooperation with the pipeline industry associations defining the agency's security program recommendations and implementation expectations. This guidance recommended that operators identify critical facilities, develop

---

[39] Colonial Pipeline, "Media Statement Update: Colonial Pipeline System Disruption," May 17, 2021, https://www.colpipe.com/news/press-releases/media-statement-colonial-pipeline-system-disruption.

[40] PDD-63, Protecting the Nation's Critical Infrastructures, May 22, 1998.

[41] In November 2004, the President signed the Norman Y. Mineta Research and Special Programs Improvement Act (P.L. 108-426), which eliminated the Research and Special Programs Administration (RSPA) and placed OPS within the newly established PHMSA. This administrative restructuring did not significantly affect the authorities or activities of OPS.

[42] 49 C.F.R. §195.436, "Security of Facilities."

[43] 49 C.F.R. §§192.613, 192.614, 192.705, 193.2509.

security plans consistent with prior trade association security guidance, implement these plans, and review them annually.[44] While the guidance was voluntary, OPS expected compliance and informed operators of its intent to begin reviewing security programs and to test their effectiveness.[45]

PHMSA Cooperation with TSA

In November 2001, President Bush signed the Aviation and Transportation Security Act (P.L. 107- 71) establishing the Transportation Security Administration (TSA) within DOT. According to TSA, the act placed DOT's pipeline security authority (under PDD-63) within TSA. The act specified for TSA a range of duties and powers related to general transportation security, such as intelligence management, threat assessment, mitigation, security measure oversight, and enforcement. On November 25, 2002, President Bush signed the Homeland Security Act of 2002 (P.L. 107-296) creating the Department of Homeland Security (DHS). Among other provisions, the act transferred TSA from DOT to DHS (§403). On December 17, 2003, President Bush issued Homeland Security Presidential Directive 7 (HSPD-7), clarifying executive agency responsibilities for identifying, prioritizing, and protecting critical infrastructure.[46] HSPD-7 maintained DHS as the lead agency for pipeline security (paragraph 15) and instructed DOT to "collaborate in regulating the transportation of hazardous materials by all modes (including pipelines)" (paragraph 22h).

In 2004, the DOT and DHS entered into an MOU concerning their respective security roles in all modes of transportation. The MOU notes that DHS has the primary responsibility for transportation security with support from the DOT and establishes a general framework for cooperation and coordination. The MOU states that "specific tasks and areas of responsibility that are appropriate for cooperation will be documented in annexes ... individually approved and signed by appropriate representatives of DHS and DOT."[47] On August 9, 2006, the departments signed an annex "to delineate clear lines of authority and responsibility and promote communications,

---

[44] James K. O'Steen, RSPA, Implementation of RSPA Security Guidance, presentation to the National Association of Regulatory Utility Commissioners, February 25, 2003.

[45] PHMSA, "Briefing: Addressing Pipeline Security Issues," https://primis.phmsa.dot.gov/comm/pipelinesecurityissuesbrief.htm.

[46] HSPD-7 supersedes PDD-63 (paragraph 37).

[47] DHS and DOT, "Memorandum of Understanding Between the Department of Homeland Security and the Department of Transportation on Roles and Responsibilities," September 28, 2004, p. 4.

efficiency, and nonduplication of effort through cooperation and collaboration between the parties in the area of transportation security."[48]

In January 2007, the PHMSA administrator testified before Congress that the agency had established a joint working group with TSA "to improve interagency coordination on transportation security and safety matters, and to develop and advance plans for improving transportation security," presumably including pipeline security.[49] According to TSA, the working group developed a multiyear action plan specifically delineating roles, responsibilities, resources, and actions to execute 11 program elements: identification of critical infrastructure/key resources and risk assessments, strategic planning, developing regulations and guidelines, conducting inspections and enforcement, providing technical support, sharing information during emergencies, communications, stakeholder relations, research and development, legislative matters, and budgeting.[50]

Clarifying PHMSA and TSA Security Roles

P.L. 109-468 required the DOT inspector general (IG) to assess the pipeline security actions taken by the DOT in implementing its 2004 MOU with the DHS (§23). The IG published this assessment in May 2008. The IG report stated

> PHMSA and TSA have taken initial steps toward formulating an action plan to implement the provisions of the pipeline security annex.... However, further actions need to be taken with a sense of urgency because the current situation is far from an "end state" for enhancing the security of the Nation's pipelines.[51]

The report recommended that PHMSA and TSA finalize and execute their security annex action plan, clarify their respective roles, and jointly develop a pipeline security strategy that maximizes the effectiveness of their respective capabilities and efforts.[52] According to TSA, working with PHMSA

---

[48] TSA and PHMSA, "Transportation Security Administration and Pipelines and Hazardous Materials Safety Administration Cooperation on Pipelines and Hazardous Materials Transportation Security," August 9, 2006.

[49] T. J. Barrett, Administrator, PHMSA, testimony before the Senate Committee on Commerce, Science, and Transportation hearing on Federal Efforts for Rail and Surface Transportation Security, January 18, 2007.

[50] Jack Fox, TSA, Pipeline Security Division, personal communication, July 6, 2007.

[51] DOT, Office of Inspector General, Actions Needed to Enhance Pipeline Security, Pipeline and Hazardous Materials Safety Administration, Report No. AV-2008-053, May 21, 2008, p. 3.

[52] Ibid., pp. 5-6.

"improved drastically" after the release of the IG report; the two agencies began to maintain daily contact, share information in a timely manner, and collaborate on security guidelines and incident response planning.[53] Consistent with this assertion, in March 2010, TSA published a Pipeline Security and Incident Recovery Protocol Plan, which lays out in detail the separate and cooperative responsibilities of the two agencies with respect to a pipeline security incident. Among other notes, the plan states

> DOT has statutory tools that may be useful during a security incident, such as special permits, safety orders, and corrective action orders. DOT/PHMSA also has access to the Regional Emergency Transportation Coordinator (RETCO) Program.... Each RETCO manages regional DOT emergency preparedness and response activities in the assigned region on behalf of the Secretary of Transportation.[54]

The plan also refers to the establishment of an Interagency Threat Coordination Committee established by TSA and PHMSA to organize and communicate developing threat information among federal agencies that may have responsibility for pipeline incident response.[55]

DOT has continued to cooperate with TSA on pipeline security in recent years. For example, TSA coordinated with DOT and other agencies to address ongoing vandalism and sabotage against critical pipelines by environmental activists in 2016.[56] In April 2016, the director of TSA's Surface Division testified about her agency's relationship with DOT:

> TSA and DOT co-chair the Pipeline Government Coordinating Council to facilitate information sharing and coordinate on activities including security assessments, training, and exercises. TSA and DOT's Pipeline and Hazardous Materials Safety Administration (PHMSA) work together to integrate pipeline safety and security priorities, as measures installed by pipeline owners and operators often benefit both safety and security.[57]

---

[53] Jack Fox, TSA, personal communication, February 2, 2010.

[54] TSA, Pipeline Security and Incident Recovery Protocol Plan, March 2010, p. 7.

[55] TSA, March 2010, p. 20.

[56] GAO, *Critical Infrastructure Protection: Actions Needed to Address Significant Weaknesses in TSA's Pipeline Security Program Management*, GAO-19-48, December 2018, p. 23.

[57] Sonya Proctor, Surface Division Director, TSA, testimony before the House Committee on Homeland Security, Subcommittee on Transportation Security hearing on "Pipelines: Securing the Veins of the American Economy," April 19, 2016.

In December 2016, PHMSA issued an Advisory Bulletin "in coordination with" TSA regarding cybersecurity threats to pipeline Supervisory Control and Data Acquisition systems.[58] In July 2017, the two agencies collaborated on a web-based portal to facilitate sharing sensitive but unclassified incident information among federal agencies with pipeline responsibilities.[59] In February 2018, the director of TSA's Surface Division again testified about cooperation with PHMSA, stating, "TSA works closely with [PHMSA] for incident response and monitoring of pipeline systems," although she did not provide specific examples.[60]

In June 2019, the Government Accountability Office (GAO) published a report examining the relative roles and responsibilities of DOT and DHS in pipeline security.[61] GAO concluded that, while the 2006 TSA-PHMSA MOU Annex delineated the agencies' mutually agreed-upon roles and responsibilities, it had not been reviewed to consider pipeline security developments since its inception. TSA's Pipeline Security and Incident Recovery Protocol Plan likewise had not been updated since it was issued in 2010 "to reflect changes in pipeline security threats, technology, federal law and policy, and any other factors."[62] Among other things, GAO recommended that TSA and PHMSA update these documents and put in place formal processes to periodically update them in the future. In response to this recommendation, TSA and PHMSA signed an update to the MOU Annex in February 2020.[63]

---

[58] PHMSA, "Pipeline Safety: Safeguarding and Securing Pipelines from Unauthorized Access," 81 *Federal Register* 89183, December 9, 2016.

[59] GAO, *Critical Infrastructure Protection: Actions Needed to Address Significant Weaknesses in TSA's Pipeline Security Program Management*, p. 23.

[60] Sonya Proctor, Surface Division Director, TSA, testimony before the House Committee on Homeland Security Subcommittee on Transportation and Maritime Security and Subcommittee on Cybersecurity, Infrastructure Protection, and Innovation joint hearing on "Securing U.S. Surface Transportation from Cyber Attacks," February 26, 2019.

[61] GAO, Critical Infrastructure Protection: Key Pipeline Security Documents Need to Reflect Current Operating Environment, GAO-19-426, June 2019.

[62] GAO, *Critical Infrastructure Protection: Key Pipeline Security Documents Need to Reflect Current Operating Environment*, pp. 29-30.

[63] PHMSA and TSA, "Annex to the Memorandum of Understanding Between the Department of Homeland Security and the Department of Transportation Concerning Transportation Security Administration and Pipeline and Hazardous Materials Safety Administration Cooperation on Pipeline Transportation Security and Safety," February 26, 2020, https://www.phmsa.dot.gov/sites/phmsa.dot.gov/files/docs/regulatory-compliance/phmsa-guidance/73466/phmsa-tsa- mou-annexexecuted.pdf.

Colonial Pipeline Incident

Following the Colonial Pipeline ransomware attack, PHMSA joined TSA and the Cybersecurity and Infrastructure Security Agency (CISA) on a teleconference call with pipeline operators to provide updates on the incident, answer questions, and provide resources to support cybersecurity mitigation efforts.[64] The Deputy Secretary of Transportation subsequently testified that PHMSA intended to "leverage its authorities to inspect and enforce three critical components of pipeline operations" related to cybersecurity: system control room regulations, integrity management plan requirements,[65] and emergency response plan regulations.[66] The Deputy Secretary also stated that DOT's Office of Intelligence, Security, and Emergency Response was collaborating with the National Security Council and interagency partners on a natural gas pipelines Industrial Control Systems Cybersecurity Initiative and that "DOT continues work with [its] sister agencies, especially TSA and CISA, to invest in world class research and pursue initiatives to address cybersecurity threats."[67]

Policy Issues for Congress

PHMSA's pipeline safety program is authorized through FY2023. In considering reauthorization, Congress may focus on oversight of the agency's ongoing regulatory activities and implementation of recent legislative priorities. Among these issues, several may be of particular interest: staffing resources; pipeline modernization; new regulation of gas gathering lines; regulation of methane leaks; PHMSA's role in pipeline security; updates to regulation of LNG, carbon dioxide, and hydrogen infrastructure; Special Permits; and pipeline safety R&D. These issues are discussed in the following sections.

---

[64] TSA, "TSA Response to Congressional Research Service Inquiry on Colonial Pipeline Incident," memorandum, June 29, 2021. Congress created CISA in the Cybersecurity and Infrastructure Security Agency Act of 2018 (P.L. 115-278). However, predecessor organizations executed similar authorities and capabilities.

[65] "An integrity management program is a set of safety management, operations, maintenance, evaluation, and assessment processes that are implemented in an integrated and rigorous manner to ensure operators provide enhanced protection" for high consequence areas. See PHMSA, "Overview: Integrity Management," https://primis.phmsa.dot.gov/comm/Im.htm.

[66] Polly Trottenberg, Deputy Secretary of Transportation, written testimony submitted for the Senate Committee on Commerce, Science, and Transportation hearing on "Pipeline Cybersecurity: Protecting Critical Infrastructure," July 27, 2021, p. 3.

[67] Ibid., pp. 4-5.

Staffing Resources for Pipeline Safety

The U.S. pipeline safety program employs a combination of federal and state staff to implement and enforce federal pipeline safety regulations. To date, PHMSA has relied heavily on state agencies for pipeline inspections, with approximately 74% of inspectors being state employees. As the PHMSA administrator remarked in 2018 PHMSA faces a manpower issue. It is obvious that [PHMSA] … cannot oversee 2.7 million miles of pipeline all by itself. In fact, PHMSA makes no attempt to do so. Most actual safety inspections are performed by our state partners.[68]

Nonetheless, some in Congress have criticized staffing at PHMSA for being insufficient to inspect pipelines under the agency's jurisdiction and to revise its regulations in line with legislative mandates and deadlines. In considering PHMSA staff levels, issues of particular interest have been the number of federal inspectors and the agency's historical use of staff funding.

In FY2023, PHMSA was funded for 356 FTE employees in pipeline safety. This total included eight new FTE positions required by the PIPES Act (§102) "to finalize outstanding rulemakings and fulfill congressional mandates." The President's requested budget authority for PHMSA's pipeline safety program in FY2024 would fund 367.5 FTE staff.

(Full-Time Equivalent Staff)

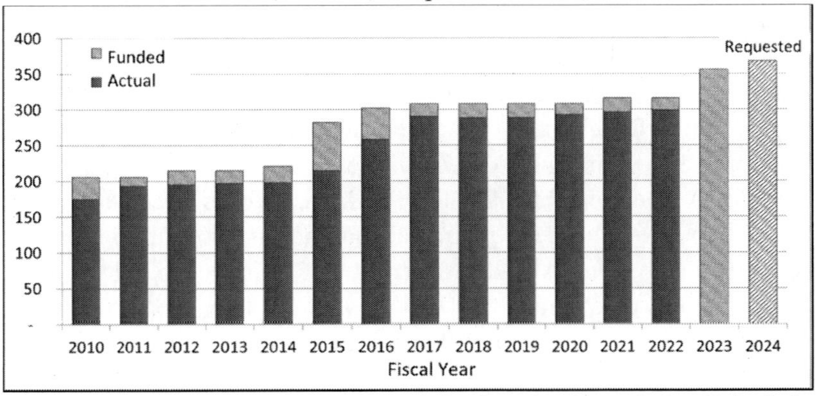

Sources: U.S. Office of Management and Budget, Budget of the United States Government: Appendix, Fiscal Years 2010-2024, "Pipeline Safety," line 1001, "Direct civilian full-time equivalent employment."

[68] Howard "Skip" Elliott, PHMSA Administrator, remarks to the Fall Pipeline Leadership Meeting of the Association of Oil Pipe Lines and the American Petroleum Institute, October 25, 2018, p. 3, https://www.phmsa.dot.gov/sites/phmsa.dot.gov/files/docs/news/69671/ aopl-api-speech.pdf.

Notes: These figures assume all staff are full-time equivalent employees (FTEs). Funded staff are "estimated staff" anticipated by the agency as reported in annual budget requests. They differ from actual staff employed (for the same fiscal year) as reported in subsequent budget requests. Pipeline safety FTEs in the figure include pipeline safety positions reporting directly through the Office of Pipeline Safety and through other program offices. PHMSA has not publicly reported actual FTEs for FY2023 as of March 31, 2023.

Figure 5. PHMSA Pipeline Safety Staffing, Historical and Requested.

As Figure 5 shows, PHMSA has faced a persistent staffing shortfall, which has generally been due to a shortage of inspectors. Agency officials have offered a number of reasons for the shortfall, including a scarcity of qualified inspector job applicants, delays in the federal hiring process (during which applicants accept other job offers), and PHMSA inspector turnover—especially due to retirements and departures to pipeline companies. Because PHMSA pipeline inspectors are extensively trained by the agency—typically for two years before being allowed to operate independently—they are highly valued by pipeline operators seeking to comply with federal safety regulations.

A 2017 DOT IG report supported PHMSA's assertions about industry-specific hiring challenges and confirmed "a significant gap between private industry and Federal salaries for the types of engineers PHMSA hires."[69] PHMSA has continued to experience staff losses due to an aging workforce and continued difficulty hiring and retaining engineers and technical staff because of competition from the oil and natural gas industry as well as workforce challenges related to the COVID-19 pandemic.[70]

Although PHMSA has acted in recent years to shore up its workforce, there have been recommendations for improvement. A 2018 GAO study stated that PHMSA had not "planned for future workforce needs for interstate pipeline inspections" and, in particular, had not assessed the resources and benefits available from its state partners.[71] GAO concluded that without this type of forward-looking analysis, PHMSA could not "proactively plan for future inspection needs to ensure that federal and state resources are in place

[69] DOT, Office of Inspector General, "PHMSA Has Improved Its Workforce Management but Planning, Hiring, and Retention Challenges Remain," Report No. ST2018010, November 21, 2017, p. 12. Congress mandated the IG study in P.L. 114-183 (§9(a)).

[70] Tristan Brown, PHMSA Acting Administrator, Keynote address at the Pipeline Safety Trust annual conference, *The Future of Pipeline Safety*, New Orleans, LA, December 2, 2022, https://youtu.be/7lo1Nu6rGe4.

[71] GAO, *Interstate Pipeline Inspections: Additional Planning Could Help DOT Determine Appropriate Level of State Participation*, GAO-18-461, May 2018, p. 16. Congress mandated the IG study in P.L. 114-183 (§24).

to provide effective oversight of interstate pipelines."[72] According to GAO, PHMSA concurred with its recommendation to develop a workforce plan for interstate pipeline inspections.

The PIPES Act (§102(b)) establishes a yearly minimum number of FTEs for pipeline safety inspection and enforcement for FY2021-FY2023. The act also requires PHMSA to "use incentives, as necessary, to recruit and retain a qualified workforce" as permitted under Title 5 of the U.S. Code, including special pay rates, student loan repayment, tuition assistance, and retention incentives. The agency has been taking measures to address its staffing challenges, such as using Direct Hiring Authority for applicable positions; investing education programs promoting pipeline safety engineering; developing targeted recruitment and hiring strategies for key positions; extending outreach among universities and professional associations; and participating in special hiring events, among other activities.[73] PHMSA has requested approval for a special pay rate table from the Office of Personnel Management for a 35% premium for engineer inspectors in its five regional offices, and continues to implement other financial incentives, including recruitment and retention bonuses, tuition assistance, and student loan repayment.[74] What impact PHMSA's workforce actions and staff incentives may have on its ongoing staffing levels—and how they may affect the agency's ability to effectively carry out its mission—may be of interest to Congress.

Aging Pipeline Modernization

The NTSB listed the safe shipment of hazardous materials by pipeline among its 2019-2020 Most Wanted List of Transportation Safety Improvements, stating "as infrastructure ages, the risk to the public from pipeline ruptures also grows."[75] Likewise, Congress has long been concerned about the safety of older transmission pipelines—a key factor in the San Bruno accident—and in leaky and deteriorating cast iron pipe in natural gas distribution systems—at issue in Merrimack Valley.[76] Construction work in Merrimack Valley, which

[72] GAO, *Interstate Pipeline Inspections*, p. 16.

[73] DOT, *Budget Estimates Fiscal Year 2024, Pipeline and Hazardous Materials Safety Administration*, 2023, p. 35.

[74] Damon Hill, PHMSA, March 24, 2023.

[75] NTSB, "Ensure the Safe Shipment of Hazardous Materials," March 28, 2019.

[76] See, for example, U.S. Congress, House Committee on Energy and Commerce, Subcommittee on Energy, *Legislative Solutions to Make Our Nation's Pipelines Safer*, committee print, 116th

led to the release of natural gas, was part of a cast iron pipe replacement project. According to the American Gas Association and other stakeholders, antiquated cast iron pipes in natural gas distribution systems, many over 50 years old, "have long been recognized as warranting attention in terms of management, replacement and/or reconditioning."[77] Old distribution pipes have also been identified as a significant source of methane leakage, which poses safety risks and contributes to U.S. GHG emissions.[78]

Natural gas distribution system operators with antiquated pipes in their systems all have programs for their replacement, although some are constrained by costs and rate regulation. Upgrading or replacing natural gas distribution infrastructure involves substantial capital investment. According to a 2015 Department of Energy analysis, the total cost of replacing all cast iron and bare steel distribution pipes in the United States at that time would be approximately $270 billion (2015 dollars).[79] These costs, in turn, could be passed on to consumers through increased natural gas rates. They could pose particular challenges for publicly owned (e.g., municipal) gas utilities with constrained budgets and limited access to capital. Practical barriers, such as urban excavation and disruption of gas supplies, also constrain annual pipe replacement. Nonetheless, as the Department of Energy stated in a 2017 report, "many policymakers and the utilities responsible for delivering natural gas to customers broadly recognize the need to accelerate ongoing efforts to replace aging infrastructure while embracing new approaches to operations and maintenance."[80]

Although the federal role in natural gas distribution systems is limited because they are under state jurisdiction, there have been past legislative proposals in Congress to provide federal support for the replacement of old cast iron pipe.[81] Likewise the House Select Committee on the Climate Crisis majority staff report, released June 2020, concluded that Congress should "provide financial support for cities and states to eliminate methane leaks from

---

Cong., 1st sess., June 19, 2019; and Office of Senator Edward Markey, "Markey Report: Leaky Natural Gas Pipelines Costing Consumers Billions," press release, Thursday, August 1, 2013.

[77] American Gas Association, "Managing the Reduction of the Nation's Cast Iron Inventory," 2013, summary.

[78] Kathryn McKain et al., "Methane Emissions from Natural Gas Infrastructure and Use in the Urban Region of Boston, Massachusetts," *Proceedings of the National Academy of Sciences*, vol. 112, no. 7 (February 27, 2015), pp. 1941-1946.

[79] Department of Energy, *Quadrennial Energy Review*, April 2015, p. 1-4.

[80] Department of Energy, *Natural Gas Infrastructure Modernization Programs at Local Distribution Companies: Key Issues and Considerations*, January 2017, p. 5.

[81] The Pipeline Revolving Fund and Job Creation Act (S. 1209, 114th Congress) introduced by Senator Markey and two cosponsors on May 6, 2015.

natural gas distribution lines within 10 years."[82] Consistent with these efforts, IIJA authorized a Natural Gas Distribution Infrastructure Safety and Modernization Grant Program administered by PHMSA. The program provides grants to municipal or community-owned natural gas distribution utilities (excluding for-profit utilities) for the repair, rehabilitation, or replacement of some or all of their pipeline systems in order to reduce safety incidents and "avoid economic losses." IIJA appropriated a total of $1.0 billion for the program in $200 million increments annually from FY2022 to FY2026 to remain available until expended.

PHMSA announced it had begun accepting applications for the grants in May 2022.[83] According to the DOT FY2024 budget request, the agency awarded no grants in FY2022 but expects to award $392 million in grants through the end of FY2023.[84] As PHMSA's implementation of the program continues, Congress may examine its structure and effectiveness along with the industry's overall progress in addressing the safety of antiquated distribution lines.

Gathering Line Regulation

Natural gas gathering lines are pipelines that collect produced gas from wellheads and transport it to centralized collection points. The latter are usually gas processing facilities where impurities are removed and gas constituents (e.g., methane, propane) are separated into distinct products for further shipment to market. Natural gas gathering lines have historically operated in mostly rural areas at lower pressure than transmission lines and with smaller diameters—typically 20 inches or less. However, due to differences in extraction techniques, especially in shale gas production with hydraulic fracturing, newer gathering lines have been constructed up to 36 inches in diameter and operated at pressures similar to those in transmission

[82] House Select Committee on the Climate Crisis, *Solving the Climate Crisis*, majority staff report, June 2020, p. 7.

[83] PHMSA, "USDOT Begins Accepting Applications for President Biden's Bipartisan Infrastructure Law Program Designed to Improve Pipeline Safety, Reduce Gas Distribution Leaks in Communities Across the Country," press release, PHMSA 04-22, May 24, 2022.

[84] U.S. Office of Management and Budget, *Budget of the United States Government, Appendix*, Fiscal Year 2024, "Natural Gas Distribution Infrastructure Safety and Modernization Grant Program," Line 0002, "Grants," p. 941.

lines.[85] Shale gas production has also been occurring in relatively more populated areas, notably the Marcellus basin in Ohio, Pennsylvania, and West Virginia. The construction of larger gathering lines in more populous regions, together with recent gathering pipeline accidents, has raised concerns about safety risks in nearby communities.[86]

The Pipeline Safety Act of 1992 (P.L. 102-508, §109) authorized PHMSA to regulate the safety of gas gathering lines that "warrant regulation," taking account of "such factors as location, length of line from the well site, operating pressure, throughput, and the composition of the transported gas." Under these provisions, PHMSA issued a 2006 final rule defining regulated gathering line that covered less than 10% of U.S. natural gas gathering line mileage at the time.[87] The remaining gathering lines were judged to pose little risk to the public due to their physical characteristics and more remote locations.

In 2011, PHMSA published an Advance Notice of Proposed Rulemaking to begin examining, among other things, whether new regulations were needed to govern the safety of natural gas gathering lines—with specific reference to shale gas lines.[88] Continuing this rulemaking process, in 2016, PHMSA published a Notice of Proposed Rulemaking (NPRM) to modify the regulation of onshore gas gathering lines—repealing an exemption for operator reporting and extending specific regulatory requirements to certain gas gathering lines with large diameters and high operating pressures in certain locations.[89]

The PIPES Act (§112(a)) required PHMSA to finalize its rule for onshore gas gathering lines by March 27, 2021. PHMSA published its final rule in the Federal Register on November 15, 2021.[90] Among its key provisions, the rule requires operators to report incidents and file annual reports for all natural gas gathering lines to "help determine the need for future regulatory changes to address the risks to the public, property and the environment."[91] According to

---

[85] PHMSA, "Pipeline Safety: Safety of Gas Transmission and Gathering Pipelines," 81 *Federal Register* 20721, April 8, 2016, p. 20728.

[86] See, for example, *Midland Reporter-Telegram*, "Report: Explosion That Killed Girl, 3, Caused by Hole in Pipeline," September 12, 2018.

[87] PHMSA, "Gas Gathering Line Definition; Alternative Definition for Onshore Lines and New Safety Standards," 71 *Federal Register* 13289, March 15, 2006.

[88] PHMSA, "Pipeline Safety: Safety of Gas Transmission Pipelines," 76 *Federal Register* 5308, August 25, 2011, pp. 3086-53102.

[89] PHMSA, "Pipeline Safety: Safety of Gas Transmission and Gathering Pipelines," pp. 20722-20856.

[90] PHMSA, "Pipeline Safety: Safety of Gas Gathering Pipelines: Extension of Reporting Requirements, Regulation of Large, High-Pressure Lines, and Other Related Amendments," 86 *Federal Register* 217, November 15, 2021, pp.63266-63299.

[91] 86 *Federal Register* 217, p. 63268.

PHMSA's announcement, under this requirement, "there are at least 425,000 miles of onshore gas gathering lines that have not been subject to PHMSA oversight but will be after this rule takes effect."[92]

The final rule also imposes new safety requirements (e.g., for damage prevention, construction, and operation) on gathering lines that have outer diameters of 8.625 inches or greater and operate at higher stress levels or pressures, with greater requirements for lines larger than 16 inches and certain gathering lines that could directly affect homes and other structures.[93] PHMSA estimates that approximately 91,000 miles of gathering lines fall into this category.[94] Operators are required to comply with safety requirement for the larger gathering lines as of May 16, 2022, with initial annual reports due by May 15, 2023.

Pipeline stakeholder representatives participated in PHMSA's gathering line rulemaking process both as members of technical panels and as commenters on the proposed rule. While stakeholders reached a consensus on many provisions in PHMSA's final rule, some remain the subject of disagreement. In December 2021, two pipeline trade associations filed a petition with PHMSA to stay enforcement and reconsider a number of specific requirements due to disagreement with the agency's risk assessment and cost-benefit determination, arguing that PHMSA is imposing excessive and unnecessary burdens on operators.[95] Conversely, pipeline safety advocates support implementing the agency's final rule "unhindered," citing the perceived "progress" in gathering line safety and concerns about industry's potentially negative influence on PHMSA's safety regulation.[96] In April 2022, PHMSA denied the petition for reconsideration of the final rule.[97] As PHMSA's final gathering line rule is implemented, compliance among operators and the effects of the final rule on overall safety in the pipeline sector may be oversight issues for Congress.

---

[92] PHMSA, "New Federal Regulations Add More Than 400,000 Miles of 'Gas Gathering' Pipelines Under Federal Oversight," press release, November 15, 2021.

[93] 86 *Federal Register* 217, p. 63268.

[94] 86 *Federal Register* 217, p. 63292.

[95] GPA Midstream Association and American Petroleum Institute, Petition for Reconsideration of Final Rule, "Safety of Gas Gathering Pipelines: Extension of Reporting Requirements, Regulation of Large, High-Pressure Lines, and Other Related Amendments," Docket No. PHMSA-2011-0023, December 20, 2021.

[96] Pipeline Safety Trust, "Pipeline Safety Trust Denounces Petition from API and GPA Midstream to Remove Important Safety Measures," press release, December 15, 2021.

[97] *Federal Register* 86, p. 26296.

PHMSA Regulation of Methane Emissions

The Environmental Protection Agency's Greenhouse Gas Inventory lists "natural gas systems" as among the highest U.S. emissions sources of atmospheric methane, a potent GHG.[98] Within this category, studies have identified pipeline emissions—arising from leaks, maintenance blowdowns, accidents, and other releases—as a major source of fugitive methane.[99] Given national goals to reduce GHG emissions in an effort to limit climate change, some in Congress have long called for tighter regulation of pipeline methane releases to reduce the sector's GHG contribution.[100] Reflecting these views, the PIPES Act (§113) mandates that PHMSA promulgate regulations requiring natural gas pipeline operators "to conduct leak detection and repair programs … to meet the need for gas pipeline safety, as determined by the Secretary; and … to protect the environment" (emphasis added). The act similarly requires PHMSA to evaluate "protection of the environment" as a factor in its review of pipeline operators' inspection and maintenance plans (§114).

The inclusion by Congress of explicit language in the PIPES Act about protecting "the environment" is widely viewed as expanding PHMSA's traditional safety mission to include climate considerations. As PHMSA's acting administrator has stated, "we need to do all we can to prevent climate change and reducing leaks which contribute to methane emission is a critical part of that."[101] The Biden Administration has likewise cited the PIPES Act provisions as elements of a national strategy to "to tackle super-polluting methane emissions—a major contributor to climate change."[102]

The provisions in the PIPES Act (§114) are self-executing, applying directly to pipeline operators. PHMSA published an advisory bulletin in the

---

[98] Environmental Protection Agency, "Data Highlights, Inventory of U.S. Greenhouse Gas Emissions and Sinks: 1990- 2019," 430-F-21-010, p. 2.

[99] See, for example: Zachary D. Weller, Steven P. Hamburg, and Joseph C. von Fischer, "A National Estimate of Methane Leakage from Pipeline Mains in Natural Gas Local Distribution Systems," *Environmental Science and Technology*, vol. 54, no. 14 (2020), pp. 8958-8967.

[100] See for, example: House Select Committee on the Climate Crisis, June 2020, pp. 200-201; Office of Senator Edward Markey, "Markey: When We Fix Leaky Natural Gas Pipelines, We Can Save Lives and Money, Create Jobs," press release, May 6, 2015.

[101] Tristan Brown, PHMSA Acting Administrator, "Remarks of PHMSA Acting Administrator Tristan Brown Before the AOPL-API Fall Meeting," October 14, 2021, https://www.phmsa.dot.gov/news/remarks-phmsa-acting-administrator-tristan-brown-aopl-api-fall-meeting.

[102] The White House, "Biden Administration Tackles Super-Polluting Methane Emissions," January 31, 2022.

Federal Register in June 2021 reminding pipeline operators to update their inspection and maintenance plans by the statutory deadline of December 27, 2021.[103] The agency is in the process of issuing an NPRM for new pipeline leak detection and repair regulations in compliance with Section 113. PHMSA conducted public meetings in May 2021 and February 2022 to gather stakeholder perspectives on the proposed rule. The agency transmitted its rulemaking package to the Office of Management and Budget for review in February 2023 with the expectation of publishing its NPRM in the Federal Register on May 15, 2023.[104]

Given PHMSA's mandate to incorporate new environmental considerations in its pipeline safety standards, its Section 114 enforcement and Section 113 rulemaking are of great interest among industry and environmental stakeholders as well as in Congress. As PHMSA implements the expanded environmental protection provisions in the PIPES Act, Congress may examine how the agency quantifies the costs and benefits of climate-related regulatory requirements, potential impacts to pipeline operations, how new information on methane leaks can inform future regulation, and how new technologies could improve leak identification and mitigation.

PHMSA and Pipeline Security

Ongoing physical and cyber threats against the nation's pipelines have heightened concerns about pipeline security risks. In a December 2018 study, GAO stated that, since the terrorist attacks of September 11, 2001, "new threats to the nation's pipeline systems have evolved to include sabotage by environmental activists and cyber attack or intrusion by nations."[105] The 2021 ransomware attack on the Colonial Pipeline Company brought pipeline security to the fore.

Recent oversight of federal pipeline safety and security activities has included discussion of PHMSA's role in pipeline security.

---

[103] PHMSA, "Pipeline Safety: Statutory Mandate to Update Inspection and Maintenance Plans to Address Eliminating Hazardous Leaks and Minimizing Releases of Natural Gas From Pipeline Facilities," 86 *Federal Register* 110, June 10, 2021, pp. 31002-31003.

[104] PHMSA, "PIPES Act 2020 Web Chart," February 24, 2023, https://www.phmsa.dot.gov/sites/phmsa.dot.gov/files/2023-02/2023%20February%20PIPES%20Act%20Chart.pdf.

[105] GAO, Critical Infrastructure Protection: Actions Needed to Address Significant Weaknesses in TSA's Pipeline Security Program Management, p. 1.

In October 2021, the PHMSA acting administrator stated that the agency's security role "includes coordination efforts with [TSA] and other federal agencies to ensure there is a collaborative and efficient approach to monitoring, inspecting, and promulgating regulations related to cybersecurity in the pipeline industry."[106] While PHMSA reports cooperation with TSA in pipeline security under the terms of the pipeline security annex and subsequent collaboration, questions may remain regarding exactly what this cooperation entails and the ongoing roles of the two agencies. Some in Congress are interested in PHMA's role in the overall federal regulatory structure overseeing pipeline security, particularly cybersecurity, and incident response.[107] In March 2023 testimony before Congress, the PHMSA Deputy Administrator stated,

> We work very closely with [TSA] There are operational impacts, potentially, when you have a cyberattack, and we're responsible for [overseeing] safe operations We've provided our input to the Transportation Security Administration on their proposed security directives on cybersecurity and we've engaged with leadership of pipeline companies to make sure we're all on the same page.[108]

In the 117th Congress, the Pipeline and LNG Facility Cybersecurity Preparedness Act (H.R. 3078) would have required the Secretary of Energy to enhance coordination among "appropriate Federal agencies," state government agencies, and the energy sector in pipeline security; coordinate incident response and recovery; support the development of pipeline cybersecurity applications, technologies, demonstration projects, and training curricula; and provide technical tools for pipeline security. What role PHMSA might play in any future pipeline security initiatives, and what resources it might require to perform that role, may be a consideration for Congress.

---

[106] Tristan Brown, October 14, 2021.

[107] See, for example, U.S. Rep. Robert Menendez, remarks before the House Transportation and Infrastructure Committee, Subcommittee on Railroads, Pipelines, and Hazardous Materials hearing on Pipeline Safety: Reviewing Implementation of the PIPES Act of 2020 and Examining Future Safety Needs, March 8, 2023.

[108] Tristan Brown, Deputy Administrator, PHMSA, testimony before the House Transportation and Infrastructure Committee, Subcommittee on Railroads, Pipelines, and Hazardous Materials hearing on *Pipeline Safety: Reviewing Implementation of the PIPES Act of 2020 and Examining Future Safety Needs*, March 8, 2023.

Carbon Dioxide Pipeline Rulemaking

Carbon dioxide (CO2) pipelines are essential components of carbon capture and storage (CCS) systems, which are proposed to reduce atmospheric emissions of man-made CO2, a greenhouse gas.[109] Pipelines are needed to transport the CO2 from where it is captured (e.g., power plants) to the underground geologic formations where it can be stored. Approximately 5,000 miles of pipeline already carry CO2 in the United States, primarily linking natural CO2 sources to aging oil fields where the CO2 is used for enhanced oil recovery.[110] However, a larger CO2 pipeline network would be needed for CCS to meet national goals for greenhouse gas reduction. One recent study suggests that such a network could total some 66,000 miles of pipeline by 2050, requiring some $170 billion in new capital investment.[111] Because CO2 in high concentrations can be hazardous to human health, building out a national CO2 pipeline network raises safety issues which may affect nearby communities and may hinder CCS deployment.

CO2 occurs naturally in the atmosphere and is produced by the human body, so it is often perceived to be relatively harmless. However, as concentrations increase, CO2 displaces oxygen—which may cause a range of negative health impacts, including suffocation.[112] Pipeline CO2 also may contain potentially hazardous contaminants, such as hydrogen sulfide. Because CO2 is colorless, odorless, and heavier than air, an uncontrolled release may spread undetected near the ground or in confined spaces. Therefore, CO2 pipelines pose a public safety risk, as demonstrated by the 2020 CO2 pipeline rupture in Satartia, MS, which led to a local evacuation and caused 45 people to be hospitalized.[113]

PHMSA has promulgated and enforces regulations for the construction, operation and maintenance, and emergency response planning for CO2

---

[109] For more information on CCS see CRS Report R44902, *Carbon Capture and Sequestration (CCS) in the United States*, by Angela C. Jones and Ashley J. Lawson.

[110] PHMSA, "Annual Report Mileage for Hazardous Liquid or Carbon Dioxide Systems," web table, March 1, 2023.

[111] Princeton University and High Meadows Environmental Institute, *Net-Zero America*, final report summary, October 29, 2021, https://netzeroamerica.princeton.edu/img/ Princeton%20NZA%20FINAL%20REPORT%20SUMMARY%20(29Oct2021).pdf.

[112] U.S. Department of Agriculture, Food Safety and Inspection Service, "Carbon Dioxide Health Hazard Information Sheet," ESHG-Health-02.00, Feb. 7, 2018.

[113] PHMSA, "Failure Investigation Report—Denbury Gulf Coast Pipelines, LLC—Pipeline Rupture/Natural Force Damage," May 26, 2022, https://www.phmsa.dot.gov/ sites/phmsa.dot.gov/files/2022-05/Failure%20Investigation%20Report%20-%20Denbury%20Gulf%20Coast%20Pipeline.pdf.

pipelines.[114] Although CO2 is listed as a Class 2.2 (non-flammable gas) hazardous material under DOT regulations, PHMSA currently applies safety requirements to CO2 pipelines similar to those for pipelines carrying hazardous liquids such as crude oil and anhydrous ammonia.[115] Prior to the Satartia accident, according to PHMSA statistics, CO2 pipeline operators reported only one injury and no fatalities caused by regulated pipelines over the last 20 years. However, pipeline safety advocates have argued that PHMSA's regulations for CO2 pipelines are insufficient with respect to hazard zones around CO2 releases, potential pipeline fractures, and corrosion of CO2 pipeline steel, among other things.[116] In response to these criticisms and findings from its own Satartia accident investigation, PHMSA announced a rulemaking on May 26, 2022, to update its CO_2 pipeline safety standards and a research solicitation to study the impact of CO_2 pipeline releases.[117]

Concerns about CO_2 pipeline safety have emerged as an issue for proposed CCS projects, especially in the Upper Midwest, where over 3,000 miles of CO_2 pipeline are in development.[118] These pipelines face opposition among affected landowners and advocacy groups for reasons including risks to public safety. As a consequence, the developers reportedly have faced resistance securing voluntary agreements with landowners for pipeline rights-of-way through their properties and there have been regulatory interventions and legislative efforts to limit state eminent domain authority for such projects.[119]

As with other federal initiatives to promote CCS deployment, Congress has acted to facilitate the construction of regional CO_2 pipeline networks

---

[114] 49 C.F.R. §§190, 195-199.

[115] 49 C.F.R. §172.101.

[116] Richard B. Kuprewicz, President, Accufacts Inc., *Accufacts' Perspectives on the State of Federal Carbon Dioxide Transmission Pipeline Safety Regulations as It Relates to Carbon Capture, Utilization, and Sequestration Within the U.S.*, prepared for the Pipeline Safety Trust, March 23, 2022, pp. 12-13.

[117] PHMSA, "PHMSA Announces New Safety Measures to Protect Americans From Carbon Dioxide Pipeline Failures After Satartia, MS Leak," press release, PHMSA 05-22, May 26, 2022.

[118] Summit Carbon Solutions, "Project Footprint," web page, March 16, 2023, https://summitcarbonsolutions.com/ project-footprint/; Navigator CO2 Ventures, "Navigator CO2," fact sheet, August 15, 2022, https://d3o151.p3cdn1.secureserver.net/wp-content/uploads/2022/08/HG-Fact-Sheet-vFINAL.pdf; Wolf Carbon Solutions, "Wolf Carbon Solutions Files Mt. Simon Hub Permit Application in Iowa," press release, February 23, 2023, https://wolfcarbonsolutions.com/wp-content/uploads/2023/02/Iowa-Permit-Release-02232023.pdf.

[119] Donnelle Eller, "Iowa Poll: Strong Majority Opposes Using Eminent Domain for Carbon-Capture Pipelines," *Des Moines Register*, March 14, 2023.

through provisions in the IIJA and the USE IT Act.[120] These acts and other legislative proposals deal primarily with financial and administrative issues, however, rather than CO_2 pipeline safety. Given the fundamental need for pipelines in CCS systems, actual or perceived safety risks associated with CO_2 pipelines may limit the potential of CCS as a greenhouse gas mitigation option. Opposition to siting of pipelines due to safety concerns may prevent CO_2 pipeline development in certain localities and increase development time and costs in others. Some advocates have suggested that Congress take a more active role by directing federal agencies to develop safety regulations specifically tailored to the distinct characteristics of CO_2 pipelines. The PHMSA Deputy Administrator testified in March 2023 that the agency planned to promulgate a draft rule "in the coming months" to set a safety standard for new CO_2 pipeline projects.[121] How PHMSA will update its CO_2 pipeline safety standards remains to be seen, but CO_2 pipeline safety, and its implications for CCS deployment, may be an oversight issue for Congress.

Hydrogen Pipeline Safety

Some in Congress have proposed hydrogen as an environmentally friendlier alternative to conventional fossil fuels for vehicles, vessels, and electric power generation. IIJA authorized an $8 billion program of Regional Clean Hydrogen Hubs, which would be centers of activity involving hydrogen production, delivery, and end use.[122] Supplying hydrogen from sources like regional hubs to power plants, industrial facilities, and vehicular fuel distribution centers could require the development of an expansive hydrogen pipeline network.

Shipping hydrogen by pipeline in the United States is not new, but the existing pipeline network is relatively small and located almost entirely along the Gulf Coast. As of March 2023, there were approximately 1,600 miles of

[120] The USE IT Act (Section 102 of Division S of P.L. 116-260) clarified CO2 pipeline eligibility for streamlined review of any necessary federal permits (e.g., for federal lands) which might be required and directed the Council on Environmental Quality to set guidance to expedite CO2 pipeline development. The Infrastructure Investment and Jobs Act (P.L. 117-58) established a Carbon Dioxide Transportation Infrastructure Finance and Innovation (CIFIA) program for CO2 pipelines and authorizes $2.1 billion over five years for low-interest CIFIA loans and grants.

[121] Tristan Brown, March 8, 2023.

[122] For more information see CRS Report R47487, *The Hydrogen Economy: Putting the Pieces Together*, by Martin C. Offutt.

active hydrogen pipeline in the United States with over 90% located in Texas, Louisiana, and Alabama, primarily serving refineries and ammonia plants.[123] The pipeline network required to support a hydrogen-based U.S. energy strategy would need to be much larger. To facilitate the pipeline transportation of hydrogen, some in Congress, in the pipeline industry, and in the executive branch have proposed blending significant hydrogen volumes with methane in existing natural gas pipelines.[124]

Transporting hydrogen by pipeline, especially in existing natural gas pipelines, poses safety challenges due to hydrogen's chemical characteristics. Hydrogen molecules are the smallest of all molecules, and therefore are more prone than methane (the principal component of natural gas) to leak through joints, microscopic cracks, and seals in pipelines and associated infrastructure.[125] Hydrogen can also permeate directly through polymer (plastic) materials, such as those typically used to make natural gas distribution pipes. The presence of hydrogen can deteriorate steel pipe, pipe welds, valves, and fittings through a variety of mechanisms, particularly embrittlement.[126] In 2022, a safety advocacy group published a report on hydrogen blending which "identifies serious concerns about the pursuit of hydrogen blending options for existing gas transmission or gas distribution pipelines" due to the potential for pipeline leaks and failures and the greater flammability of hydrogen compared to methane.[127] However, a pipeline

---

[123] PHMSA, "Gas Distribution, Gas Gathering, Gas Transmission, Hazardous Liquids, Liquefied Natural Gas (LNG), and Underground Natural Gas Storage (UNGS) Annual Report Data," Form 7100.2-1 operator filings database, 2023, available at https://www.phmsa.dot.gov/data-and-statistics/pipeline/gas-distribution-gas-gathering-gas-transmission-hazardous-liquids. The other states with hydrogen pipelines are Kansas, Michigan, New York, Ohio, Oklahoma, Utah, and Washington.

[124] See, for example, U.S. Senator Joseph Manchin, opening remarks before the U.S. Senate Committee on Energy and Natural Resources hearing on *Opportunities and Challenges in Using Clean Hydrogen in the Transportation, Utility, Industrial, Commercial, and Residential Sectors*, February 10, 2022; Kavya Balaraman, "SoCalGas, SDG&E Outline Plan for Hydrogen Blending Demonstration Projects in California," September 20, 2022, *Utility Dive*, https://www.utilitydive.com/news/socalgas-sdge-hydrogen-blending-cpuc/632 201/.

[125] The kinetic diameters of molecular hydrogen and methane, respectively, are 289 and 380 picometers.

[126] Peter Adam et al., "Hydrogen Infrastructure—The Pillar of Energy Transition," white paper, Siemens Energy, September 15, 2020, pp. 14-15, https://assets.siemens-energy.com/siemens/assets/api/uuid:3d4339dc-434e-4692-81a0- a55adbcaa92e/200915-whitepaper-h2-infrastructure-en.pdf.

[127] Richard B. Kuprewicz, President, Accufacts Inc., *Safety of Hydrogen Transportation by Gas Pipelines*, prepared for the Pipeline Safety Trust, November 28, 2022, p. 1, https://pstrust.org/wp-content/uploads/2022/11/11-28-22-Final-Accufacts-Hydrogen-Pipeline-Report.pdf.

industry trade group disagreed with these findings, pointing to operator experience safely transporting hydrogen blends.[128]

PHMSA's pipeline safety authority extends to hydrogen pipelines, which the agency has regulated since 1970 as a "flammable gas."[129] PHMSA does not currently prohibit natural gas pipeline operators from introducing hydrogen in their systems. However, some stakeholders have questioned whether PHMSA's existing regulations are appropriate and sufficient to ensure the safety of an expanding hydrogen pipeline network, especially if it includes existing natural gas pipelines carrying hydrogen blends.[130] For example, the agency does not currently require operators to report information about hydrogen blends in their pipeline systems if natural gas is the dominant commodity. A 2021 Sandia National Laboratories report reviewing pipeline industry standards concluded,

> There are many safety codes and standards that are relevant to hydrogen blending in the natural gas infrastructure. Relevant codes include those that address natural gas and hydrogen specifically, as well as those that address blended gasses. However, there are gaps that will need to be addressed when considering introducing hydrogen/natural gas blends into the current infrastructure.[131]

PHMSA's research and development program database currently lists eight active projects related to hydrogen infrastructure safety funded under its existing research grant program.[132] Congress has promoted additional federal initiatives around hydrogen pipeline safety-related research and development. The IIJA directs the Secretary of Energy to advance the safe and efficient delivery of hydrogen or hydrogen-carrier fuels in pipelines, including by retrofitting existing natural gas pipelines (§40313). In the 118th Congress, the Hydrogen Infrastructure Finance and Innovation Act (S. 649) would establish

---

[128] American Gas Association and American Public Gas Association, joint letter to Bill Caram, Executive Director, Pipeline Safety Trust and Richard Kuprewicz, President, Accufacts Inc., December 15, 2022, https://pstrust.org/wp- content/uploads/2023/01/AGA-letter-to-PST-on-H-Report.pdf.

[129] PHMSA regulates hydrogen pipeline safety under its safety requirements at 49 C.F.R. Part 192, "Transportation of Natural and Other Gas by Pipeline: Minimum Federal Safety Standards."

[130] Mike Soraghan, "Biden Energy Agenda Exposes Regulatory Gap," *E&E News*, February 26, 2023.

[131] Sandia National Laboratories, Codes and Standards Assessment for Hydrogen Blends into the Natural Gas Infrastructure, SAND2021-12478, October 2021, p. 31.

[132] PHMSA, "Research and Development Program Awards," web database, accessed March 15, 2023, https://primis.phmsa.dot.gov/matrix/prjquery.rdm.

a DOE hydrogen infrastructure finance and innovation pilot program. The act also would require federal agencies, including PHMSA, to cooperatively study outstanding questions regarding research, development, and demonstration of hydrogen pipeline infrastructure and to separately study "jurisdiction over the siting, construction, safety, and regulation of hydrogen transportation infrastructure, including, at a minimum, the blending of hydrogen in natural gas pipelines." The bill would also impose hydrogen leakage, monitoring, reporting, verification, detection, and repair requirements on pipelines receiving federal financial support under the bill.

PHMSA officials have stated that the agency is "taking a look at" revising its regulations for hydrogen pipelines but that "research ... needs to be done, we need to know more" to ensure that any potential future changes to its regulations appropriately address risks to hydrogen pipeline safety.[133] Some stakeholders have questioned whether new hydrogen pipeline projects, especially blending projects, should be permitted while PHMSA's existing regulations are being reexamined. As hydrogen infrastructure research, development, and deployment continues, the adequacy of PHMSA's hydrogen pipeline safety regulation may be an issue for Congress.

Special Permits

If a pipeline operator believes unique circumstances would make it impracticable or inappropriate to comply with PHMSA's pipeline safety regulations, the operator may apply to the agency for a Special Permit to waive or modify compliance. By statute, PHMSA is authorized to "waive compliance with any part of an applicable standard ... with respect to such facility on terms the Secretary [of Transportation] considers appropriate if the Secretary determines that the waiver is not inconsistent with pipeline safety."[134] PHMSA's website lists 112 Special Permits (formerly called Waivers) for pipelines dating back to 1976.[135] Over the last four years, the

---

[133] Alan Mayberry, PHMSA Associate Administrator, remarks at the The Future of Pipeline Safety—Technology, Tools, and Transition conference, "Hydrogen Pipeline Safety" session, sponsored by the Pipeline Safety Trust, New Orleans, LA, December 2, 2022, https://www.youtube.com/watch?v=5uECL9-Gc9M.

[134] 49 U.S.C. §60118(c).

[135] PHMSA, "Special Permits Issued," web table, September 8, 2022, https://www.phmsa.dot.gov/pipeline/special- permits-state-waivers/special-permits-issued.

agency has issued 43 such permits, nearly all for natural gas transmission pipelines, and denied two permit applications.[136]

PHMSA's authorization of Special Permits became an issue of heightened concern after the December 2022 oil spill near Washington, KS, from the Keystone Pipeline, which released an estimated 14,000 barrels of crude oil and impacted Mill Creek.[137] The pipeline was operating under a Special Permit, issued in 2007, allowing it to operate at a pressure level of 80% of the pipeline's specified minimum yield strength (SMYS) rather than the standard limit of 72% of SMYS.[138] Some in Congress and other stakeholders have questioned PHMSA's issuance and enforcement of the Keystone Pipeline Special Permit, and the agency's use of such permits in general.[139] After the Keystone Pipeline spill, PHMSA reportedly commissioned Oak Ridge National Laboratory to review the agency's Special Permits program, as well as individual permits, although the agency has not publicly announced such a review.[140]

Outdated LNG Safety Standards

The adequacy of PHMSA's minimum safety standards for LNG facilities (49 C.F.R. §193) has become a concern in Congress due to growth in U.S. LNG infrastructure and recent safety incidents.[141] Although PHMSA has no siting authority, FERC requires compliance with PHMSA's regulations for the siting and operation of LNG marine terminals for import or export.[142] In August

---

[136] Ibid.; PHMSA, "Special Permits Denied," web table, April 23, 2021, https://www.phmsa.dot.gov/pipeline/special- permits-state-waivers/special-permits-denied.

[137] PHMSA, "In the Matter of TC Oil Pipeline Operations, Inc.," Corrective Action Order, CPF No. 3-2022-074-CAO, December 8, 2022, https://www.phmsa.dot.gov/sites/phmsa.dot.gov/files/2022-12/TC%20Oil%20CAO.3-2022- 074.pdf.

[138] PHMSA, Special Permit PHMSA-2006-26617, April 30, 2007, https://www.phmsa.dot.gov/sites/phmsa.dot.gov/files/docs/TC_Keystone_2007-04-30_508compliant.pdf.

139 See, for example, Senate Committee on Commerce, Science, and Transportation, "Cantwell Calls for Increased Oversight After Keystone Pipeline Spills Nearly 600,000 Gallons of Tar Sands, Largest Onshore Spill in Nearly a Decade," press release, January 9, 2023.

[140] Nia Williams and Rod Nickel, "U.S. Pipeline Regulator Reviewing Special Permits After Keystone Oil Spills,' *Reuters*, December 21, 2022.

[141] PHMSA's statutory authority for LNG is codified at 49 U.S.C. §60103.

142 The siting provisions in 49 C.F.R. §193 incorporate by reference Standard 59A, Standard for the Production, Storage, and Handling of Liquefied Natural Gas (LNG), from the National Fire Protection Association (NFPA). NFPA 59A requires thermal exclusion zones and flammable vapor-gas dispersion zones around LNG terminals (§§193.2057, 193.2059). PHMSA regulations also adopt many of NFPA's design and construction guidelines including

2020, GAO published a study of U.S. LNG exports which examined PHMSA's regulation of LNG terminal safety, among other topics. The study reported,

> PHMSA's Part 193 regulations for permitting LNG export facilities, last revised in 2015, incorporate nine technical standards that, according to a PHMSA document, are the basis for FERC's safety review of LNG export facilities. Eight of the nine incorporated standards are outdated.[143]

The PIPES Act of 2020 subsequently required PHMSA to review its minimum LNG operating and maintenance standards and, based on its review, to update its standards for "large-scale liquefied natural gas facilities (other than peak shaving facilities) to provide for a risk-based regulatory approach for such facilities" (Sec. 110a).[144] PHMSA was given a three-year deadline from enactment to complete these tasks.

The issue of outdated LNG facility standards initially drew scrutiny after the 2018 partial shutdown of the Sabine Pass LNG terminal in Cameron Parish, TX, due to cracks found in LNG storage tanks which resulted in leaking LNG.[145] This incident was followed by the June 8, 2022, accident at the Freeport LNG export terminal. In the latter incident, a piping failure caused a rapid release of methane, forming a flammable vapor cloud which subsequently exploded as a massive fireball.[146] Although no injuries were reported, the incident caused significant damage to adjacent piping, electrical systems, and other facility infrastructure. The LNG terminal was forced to shut down to make repairs and conduct safety recommissioning, temporarily halting approximately 20% of U.S. LNG exports. Freeport LNG was able to resume full operation in March 2023, approximately eight months after the

---

requirements for LNG facilities to withstand fire, wind, hydraulic forces, and erosion from LNG spills (§§193.2067, 193.2155, 193.2301).

[143] GAO, Natural Gas Exports: Updated Guidance and Regulations Could Improve Facility Permitting Processes, GAO-20-619, August 2020, p. 29.

[144] Peak shaving facilities draw natural gas from the pipeline system during periods of low demand and liquefy it into LNG for long-term storage. The LNG can then be regasified and re-injected into the pipeline system at a later time to supplement pipeline gas supplies during times of peak demand.

[145] PHMSA, "Notice of Probable Violation and Proposed Civil Penalty," CPF 4-2021-002-NOPV, Cheniere Energy, Inc., July 1, 2021.

[146] IFO Group, *Freeport LNG Quintana Island, Texas: June 8, 2022—Loss of Primary Containment Incident Investigation Report*, October 30, 2022, pp. 49-50, https://www.phmsa.dot.gov/sites/phmsa.dot.gov/files/2022-11/IFO-Group-RCFA-Report-final-redacted.pdf.

accident.[147] In the wake of the Freeport accident, pipeline safety advocates and community stakeholders called for greater urgency in updating PHMSA's LNG safety requirements.[148]

The PHMSA Deputy Administrator testified in March 2023 that the agency's updated LNG rule was a "priority" and that he "hoped to get a proposal this year."[149] According to PHMSA's public tracker for PIPES Acts rulemakings, the agency expects to publish a revised LNG safety standard by September 29, 2023.[150] Whether PHMSA promulgates its new regulations by this date, and whether the revisions to its regulations appropriately incorporate the newest industry standards to reduce LNG safety risks, may be an oversight issue for Congress.

Pipeline Safety Research, Development, and Demonstration

Congress provides PHMSA with funding for pipeline safety-related research and development. According to PHMSA, the agency "conducts and supports research to support regulatory and enforcement activities and to provide the technical and analytical foundation necessary for planning, evaluating, and implementing the pipeline safety program."[151] As of March 22, 2023, PHSMA's database lists 63 active projects supported by approximately $38.8 million in agency funding.[152] For FY2023, enacted appropriations for the agency's pipeline safety research and development (R&D) program are $12.5 million. The President's FY2024 budget would increase these appropriations to $15.0 million in 2024.[153]

In addition to R&D activities funded by PHMSA, the PIPES Act of 2020 included provisions allowing PHMSA to "establish and carry out limited safety-enhancing testing programs to evaluate innovative technologies and

[147] Freeport LNG, "Freeport LNG Receives Regulatory Approval for Restart of Final Liquefaction Train," press release, March 8, 2023, http://freeportlng.newsrouter.com/news_ release.asp?intRelease_ID=9760&intAcc_ID=77.

[148] Mike Soraghan and Mike Lee, "LNG Explosion Shines Light on 42-Year-Old Gas Rules," *E&E News*, June 28, 2022.

[149] Tristan Brown, March 8, 2023.

[150] PHMSA, "PIPES Act 2020 Web Chart," February 24, 2023.

[151] PHMSA, "PHMSA Research and Development," web page, https://www.phmsa.dot.gov/research-and-development/phmsa-research-and-development.

[152] PHMSA, "Research Project Query," web database, accessed March 22, 2023, https://primis.phmsa.dot.gov/matrix/ prjQuery.rdm.

[153] DOT, *Budget Estimates Fiscal Year 2024: Pipeline and Hazardous Materials Safety Administration*, Exhibit III-1, p. 32.

operational practices" for natural gas and hazardous liquid pipeline facilities implemented by pipeline operators (§104). In February 2002, PHMSA published in the Federal Register a notice outlining how the agency would review and process Pipeline Safety Enhancement Program (PSEP) applications by pipeline owners and operators, establishing a three-year time limit for the duration of a PSEP pilot project, and setting an application deadline of December 21, 2023.[154] However, according to March 2023 hearing testimony from pipeline industry representatives, no operator has participated in the PSEP technology pilot program because, in their view, PHMSA did not allow sufficient time for the program to operate and imposed excessive administrative and review requirements on the PSEP applications.[155] At the same hearing the PHMSA Acting Administrator testified that a high standard of safety review is necessary because PSEP projects have safety implications for the public.[156] He also testified that environmental review is mandated under the National Environmental Policy Act (NEPA) but that it is a "common goal" to make the program more efficient.[157]

Apart from PHMSA's R&D activities, Congress has funded pipeline safety-related research through DOE. For example, DOE's Hydrogen Program funded a study from 2020-2022 by Sandia National Laboratories on hydrogen blending in natural gas pipelines which examined, among other things, hydrogen-induced degradation of distribution piping and gaps in related safety codes and standards.[158] In addition, DOE's Office of Fossil Energy and Carbon Management, through its Carbon Transport program and in collaboration with PHMSA and other federal agencies, works "to ensure a safe and reliable CO2 transport network that supports the deployment of carbon capture, utilization, and storage (CCUS) and carbon dioxide removal (CDR)."[159]

---

[154] 87 *Federal Register* 22, February 2, 2022, p. 5939.

[155] Andrew Black, Chief Executive Officer, Liquid Energy Pipeline Association (LEPA), and Kenneth W. Grubb, Chief Operating Officer, Gas Pipelines, Kinder Morgan, Inc., testimony before the House Transportation and Infrastructure Committee, Subcommittee on Railroads, Pipelines, and Hazardous Materials hearing on *Pipeline Safety: Reviewing Implementation of the PIPES Act of 2020 and Examining Future Safety Needs*, March 8, 2023.

[156] Tristan Brown, March 8, 2023.

[157] Tristan Brown, October 14, 2021. NEPA is codified at 42 U.S.C. §§4321 et seq.

[158] Sandia National Laboratories, "Hydrogen Blending into Natural Gas Pipelines," Project ID: H2060, WBS 8.6.4.2, slide presentation at the DOE Hydrogen Program 2022 Annual Merit Review and Peer Evaluation Meeting, June 6, 2022.

[159] National Energy Technology Laboratory, "Carbon Transport," web page, https://netl.doe.gov/carbon-management/ carbon-storage/transport.

A discussion draft bill introduced in March 2023 in the House Committee on Science, Space, and Technology, Subcommittee on Energy, the Next Generation Pipelines Research and Development Act would require the DOE, in coordination with PHMSA, to establish a new initiative to fund

> demonstration projects on low- to mid-technology readiness level subjects ... applicable to pipelines and associated infrastructure, including liquefied natural gas facilities and underground and above ground gas and liquid fuel storage facilities; and ... development of next generation pipeline systems, components, and related technologies (§4(a)).[160]

Focus areas under the initiative would include advanced leak detection and mitigation, novel materials, technologies and methods for retrofitting existing pipelines, advanced sensors, and technologies and methods to reduce potential environmental impacts, among others. The initiative would prioritize a diverse mix of commodities, including gas and liquid hydrocarbons, carbon dioxide, hydrogen, and hydrogen blends. Another discussion draft bill introduced in March 2023 in the House Committee on Science, Space, and Technology's Subcommittee on Energy would direct DOE to carry out a hydrogen technology research, development, and demonstration program including activities focused on hydrogen pipelines, hydrogen leakage, and retrofitting of modifying "existing energy infrastructure, including existing natural gas transportation infrastructure for the purpose of transportation and storage of significant quantities of hydrogen and hydrogen blend" (§3(c)(3)(C)).[161]

As the programs above indicate, Congress has supported ongoing initiatives within PHMSA, DOE, and the pipeline industry, to develop and deploy new pipeline safety technologies and operating practices. Budgetary and legislative proposals in the 118th Congress would expand these initiatives. Ensuring that these programs are implemented and coordinated effectively

---

[160] House Committee on Science, Space, and Technology, Subcommittee on Energy, draft bill, March 23, 2023, https://republicans-science.house.gov/_cache/files/0/8/08d9c944-64a5-4b3b-9f03-4e231ff5aaa2/BC1BAA5E27E2E1EFB424D74C28B0B646.2023-03-23-discussion-draft-pipelines-rd.pdf.

[161] House Committee on Science, Space, and Technology, Subcommittee on Energy, draft bill "To direct the Secretary of Energy to conduct a program of research, development, demonstration, and commercial application with respect to clean hydrogen and fuel cell energy, low-emission fuels, and coproducts, and for other purposes," March 23, 2023, https://science.house.gov/hearings?ID=4E95E346-1854-4E0D-B991-9BA5F670D5EF.

among the various entities involved may require additional congressional oversight and direction.

Conclusion

Government and industry have taken numerous steps to improve pipeline, natural gas storage, and LNG infrastructure safety over the past 10 years. Nonetheless, major oil and natural gas pipeline accidents and security incidents continue to occur. Congress and various stakeholders have called for additional regulatory measures to reduce the likelihood of future failures. Recent PHMSA reauthorizations have included expansive pipeline safety mandates, such as requirements for the agency to regulate underground natural gas storage, significantly increase inspector staffing, and account for the climate impacts of methane leaks. Congress may consider new regulatory mandates for PHMSA or may impose new requirements directly on the pipeline industry.

However, significant changes to pipeline safety regulation are being implemented, and certain rulemakings remain outstanding, so their effects on pipeline safety have yet to be determined. The emergence of new safety risks from the development of carbon dioxide and hydrogen pipeline infrastructure raises additional regulatory challenges. As Congress continues its oversight of the federal pipeline safety program, an important focus may be the practical effects of the many changes being made to particular aspects of PHMSA's pipeline safety regulations.

In addition to the specific issues highlighted in this report, Congress may assess how the many elements of U.S. pipeline safety activity fit together in the nation's overall strategy to protect the public and the environment. Pipeline safety necessarily involves various groups: federal and state agencies, tribal governments, pipeline associations, large and small pipeline operators, local communities, and other interest groups. Reviewing how these groups work together to achieve common goals or resolve conflicting approaches could be an overarching concern for Congress.

Chapter 4

Gas Pipeline Safety: Better Data and Planning Would Improve Implementation of Regulatory Changes*

U.S. Government Accountability Office

Why This Matters

About 300,000 miles of natural gas transmission pipelines across the United States carry products from processing facilities to communities and other large- volume customers. Pipelines are a relatively safe mode for transporting natural gas, but incidents can still occur that result in death, injury, and property and environmental damage.

The Department of Transportation's Pipeline and Hazardous Materials Safety Administration (PHMSA) sets the federal minimum safety standards for these pipelines. In 2003, PHMSA established integrity management—a risk-based approach to managing certain gas transmission pipelines—as an addition to its existing pipeline safety regulations. Under this approach, operators are required to assess pipelines in high consequence areas (HCA)—generally, areas where an incident could have the greatest impact to public safety or property—to identify threats and mitigate risks. In October 2019 and August 2022, PHMSA issued final rules that both strengthened its gas transmission pipeline safety regulations and expanded some integrity

* This is an edited, reformatted and augmented version of the United States Government Accountability Office Q&A Report to Congressional Committees, Publication No. GAO-24-106690, dated April 3, 2024.

In: Pipeline Safety: Background, Practices and Regulatory Changes
Editor: Michael E. Massaro
ISBN: 979-8-89530-058-9

assessment requirements beyond HCAs, including to newly defined moderate consequence areas (MCA).

The Protecting our Infrastructure of Pipelines and Enhancing Safety Act of 2016 includes a provision for us to examine gas transmission integrity management programs following PHMSA's completion of a specific pipeline safety rulemaking.[1] We are providing information on potential changes to the methods operators use to identify HCAs; how selected stakeholders, including pipeline operators and state inspectors, view the regulatory changes to gas transmission pipeline safety; and how PHMSA is overseeing the implementation of the 2019 and 2022 final rules stemming from the rulemaking.

Key Takeaways

- PHMSA officials told us they are considering regulatory changes to the calculation of a pipeline's potential impact radius. However, when a pipeline incident occurs, PHMSA does not collect detailed data comparing the actual impact of the incident with the potential impact radius that the operator had calculated for the pipeline. Without data on the full impact of these pipeline incidents, PHMSA may find it more difficult to assess the method's accuracy.
- Selected stakeholders identified key changes in the 2019 and 2022 final rules that they expect will improve gas transmission pipeline safety by increasing the pipeline mileage assessed, improving data and information, and strengthening requirements for operators to take preventative and remedial actions.
- PHMSA has largely implemented the 2019 final rule and is in the process of implementing the 2022 rule; however, it has not created an implementation plan that includes timelines and objectives for the 2022 final rule. Without a plan, PHMSA may not be able to ensure that the agency accomplishes its remaining activities in a timely manner and conducts outreach to the appropriate audiences.
- PHMSA's Final Rule Implementation web pages do not include complete information and guidance for both rules. Without action from PHMSA to update the web pages, operators and inspectors may have difficulties when trying to find information on the rules.

[1] Pub. L. No.114-183, § 4, 130 Stat. 514, 517 (2016).

- We are recommending that PHMSA evaluate what additional data are needed from operators to better understand the actual impact of pipeline incidents, as the agency considers changes to the potential impact radius calculation. We are also recommending that PHMSA develop a plan for the remaining activities to implement the 2022 final rule and include all relevant information and guidance on its Final Rule Implementation web pages.

What Are Gas Transmission Pipelines?

Natural gas transmission pipelines carry gas, sometimes over hundreds of miles, to communities and large-volume users (e.g., factories).[2] In 2023, PHMSA officials estimated that the United States has over 300,000 miles of onshore gas transmission pipelines. Roughly two-thirds of these miles were *interstate* pipelines, or pipelines that generally cross state boundaries. The remaining third were *intrastate* pipelines, or pipelines that tend to operate within a single state.[3] Transmission pipelines tend to have larger diameters and operate at higher pressures than other types of pipelines (see fig.1).

How Frequent and What Are the Causes of Significant Gas Transmission Pipeline Incidents?

PHMSA data show that significant onshore gas transmission pipeline incidents have averaged 61 incidents per year from 2010-2022 (see fig. 2). Few of these occurred in HCAs—generally, where incidents could have the greatest impact to public safety or property. PHMSA defines a significant incident as one that results in a fatality, an injury requiring hospitalization, or property damage that exceeds $50,000 in total costs (in 1984 dollars).

---

[2] According to PHMSA data, gas pipelines in the United States primarily transport natural gas. However, other commodities such as hydrogen may also be transported via pipelines.

[3] The Pipeline Safety Laws, 49 U.S.C. § 60101 et seq., define interstate gas transmission pipelines to also be subject to the Federal Energy Regulatory Commission's (FERC) jurisdiction under the Natural Gas Act, 15 U.S.C. § 717 et seq., and intrastate gas transmission pipelines to also not be subject to FERC's jurisdiction under such act. See 49 U.S.C. § 60101(a)(6), (9).

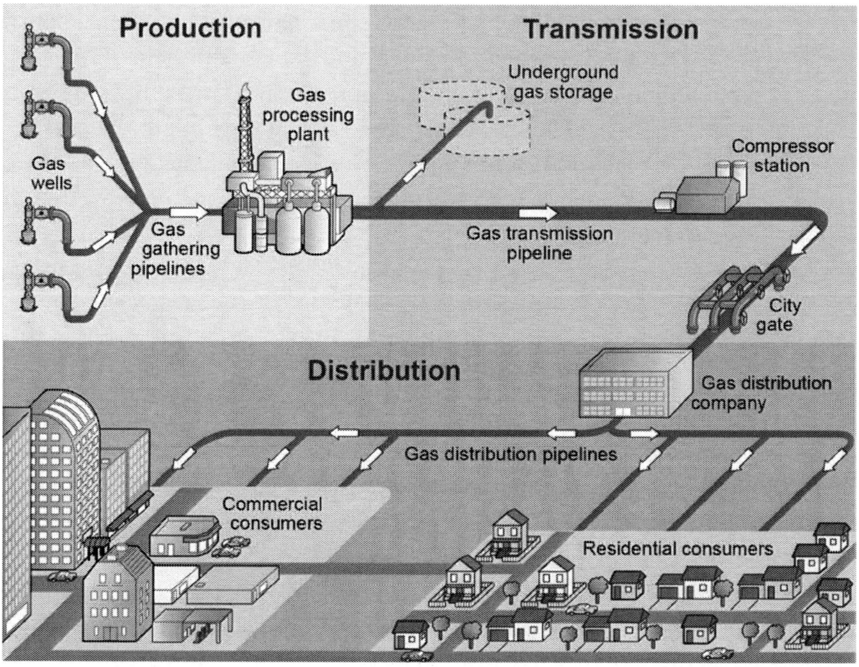

Source: GAO analysis of Energy Information Administration and Natural Gas Council documents, GAO (illustration). | GAO-24-106690.

Figure 1. Transmission Pipelines within a Natural Gas Pipeline System.

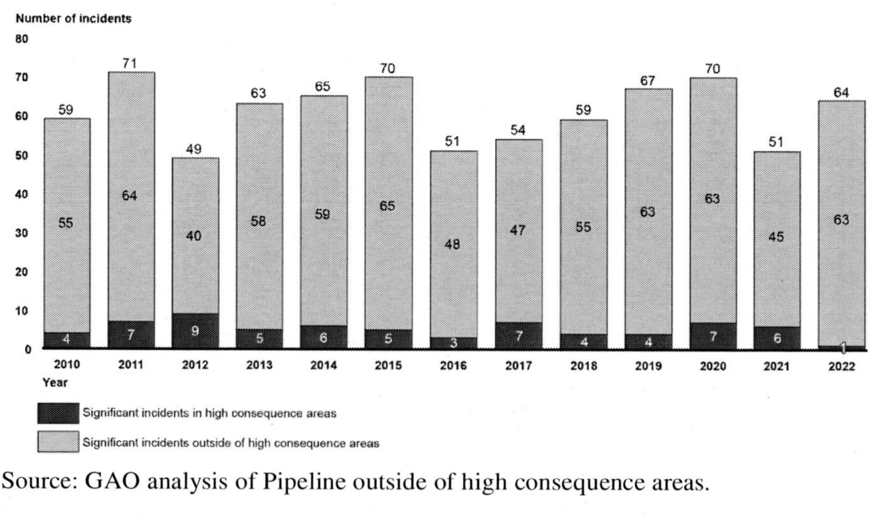

Source: GAO analysis of Pipeline outside of high consequence areas.

Figure 2. Significant Onshore Gas Transmission Pipeline Incidents Reported by Operators 2010-2022.

A variety of factors—including mechanical failure, corrosion, and excavation damage—cause pipeline incidents. From 2010-2022, PHMSA data show that nearly half (43 percent) of significant onshore gas transmission pipeline incidents were caused by equipment or material failure, followed by corrosion failure (16 percent).

Significant pipeline incidents in HCAs had slightly different causes (see fig. 3). Operators reported that approximately 37 percent of these incidents were caused by equipment or material failure, and approximately 19 percent of incidents were the result of excavation damage.

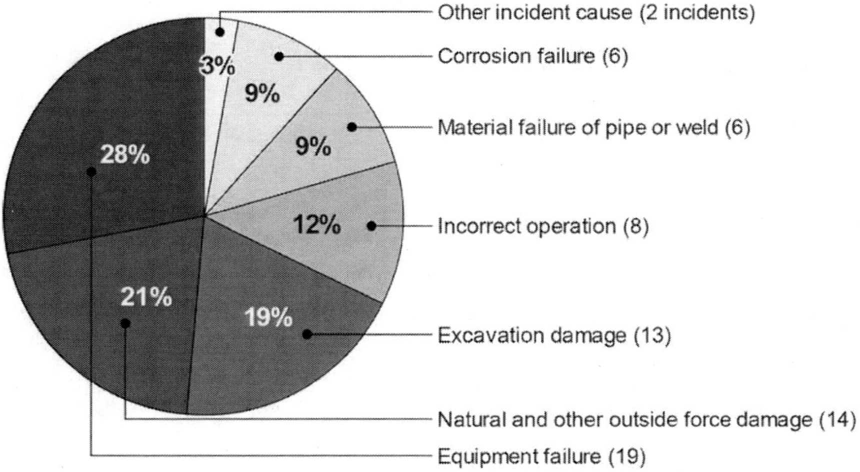

Source: GAO analysis of Pipeline and Hazardous Materials Safety Administration data.

Note: Percentages above are rounded to the nearest percentage point.

Figure 3. Reported Causes of Significant Onshore Gas Transmission Pipeline Incidents in High Consequence Areas, 2010-2022.

According to PHMSA, pipeline equipment and materials can fail due to malfunction of equipment like valves or compressors or design defects. In HCAs, excavation damage by the operator or a third party is a more common cause of a pipeline incident than in non-HCAs.

How does PHMSA Oversee Gas Transmission Pipeline Safety?

PHMSA sets the federal minimum safety standards for interstate and intrastate gas transmission pipelines, including requirements for their design, construction, testing, inspection, operation, and maintenance.[4] The agency also oversees pipeline safety through enforcing these regulations, conducting inspections, and other efforts, including requiring operators to submit annual and incident reports.[5] PHMSA conducts pipeline inspections primarily through its five regional and two district offices.

States may assume regulatory, inspection, and enforcement responsibilities for intrastate transmission pipelines within their borders, provided that the state certifies to PHMSA that it has adopted all applicable federal safety standards and is enforcing them as state laws, among other things.[6] Though PHMSA is exclusively responsible for enforcing its regulations for interstate pipelines, the agency may also authorize states holding certifications to assist with its oversight and inspections of these pipelines as "interstate agents" of PHMSA. As of 2022, eight states are authorized to act as interstate agents of PHMSA for gas pipelines.[7]

PHMSA also oversees the risk-based regulatory program known as integrity management.[8] Under PHMSA's regulations, operators with pipelines in HCAs are required to develop their own integrity management programs.[9]

---

[4] PHMSA's general authority is under the Pipeline Safety Laws. Its minimum safety standards for gas pipelines are located in 49 C.F.R. Part 192.

[5] PHMSA's reporting requirements for gas pipeline operators are located in 49 C.F.R. Part 191. These regulations require gas transmission pipeline operators to file annual reports that include information on pipeline characteristics such as location, age, and material, and submit reports on incidents. PHMSA's regulations define an incident to include any event involving a release of gas that (1) results in a fatality or injury requiring in-patient hospitalization; (2) results in estimated property damage that meets or exceeds a certain threshold; (3) results in an unintentional estimated gas loss that meets or exceeds a certain threshold; or (4) is significant in the judgment of the operator, even though the event did not meet the specific criteria under PHMSA's regulatory definition of an incident.

[6] If a federal minimum standard is issued within a certain time frame before a state submits its certification, the state must certify that it is taking steps to adopt that standard into its laws. States holding certifications may also adopt additional or more stringent requirements for these intrastate pipelines, so long as they are compatible with federal requirements. *See* 49 U.S.C. §§ 60104(c), 60105.

[7] As of 2022, the eight states authorized to act as interstate agents are Arizona, Connecticut, Iowa, Michigan, Minnesota, New York, Ohio, and Washington.

[8] PHMSA's integrity management program regulations, issued in 2003, are located in Subpart O of 49 C.F.R. Part 192. *See* Pipeline Safety: Pipeline Integrity Management in High Consequence Areas (Gas Transmission Pipelines), 68 Fed. Reg. 69778 (Dec. 15, 2003).

[9] The definition of an HCA is located in 49 C.F.R. § 192.903.

These programs must identify all HCAs, integrate relevant pipeline data, and include a risk assessment to identify potential threats to pipelines in those areas. The regulations require operators to conduct regular integrity assessments of those pipelines and correct any defects they find.

What Recent Changes Has PHMSA Made to Gas Transmission Pipeline Safety Regulations?

PHMSA recently amended its gas transmission pipeline safety regulations, including parts of its integrity management program regulations, through two final rules issued in October 2019 and August 2022.[10] The rules required operators to improve their data and records of pipeline characteristics and reconfirm the maximum allowable operating pressure of certain pipelines, and adjusted repair criteria for addressing pipeline defects in HCAs, among other things. The rules also changed or established requirements related to operators' integrity assessments of pipelines in HCAs, MCAs, and other specific areas (see fig. 4).

The 2019 final rule went into effect on July 1, 2020, with additional compliance deadlines specified for various provisions of the rule. The 2022 final rule went into effect on May 24, 2023. However, prior to the rule's effective date, PHMSA notified operators that it would not enforce most of the rule's provisions until February 24, 2024, with certain limited exceptions. The agency did this to facilitate operator compliance and implementation efforts.[11] According to officials, PHMSA used this period to make several technical clarifications to the 2022 final rule, and as a result officials expect improved operator compliance with the regulatory changes.

---

[10] Pipeline Safety: Safety of Gas Transmission Pipelines: MAOP Reconfirmation, Expansion of Assessment Requirements, and Other Related Amendments, 84 Fed. Reg. 52180 (Oct. 1, 2019); and Pipeline Safety: Safety of Gas Transmission Pipelines: Repair Criteria, Integrity Management Improvements, Cathodic Protection, Management of Change, and Other Related Amendments, 87 Fed. Reg. 52224 (Aug. 24, 2022).

[11] Exercising its enforcement discretion, PHMSA issued three notices regarding certain onshore gas transmission pipeline operators' compliance with most provisions of the 2022 final rule. *See* Notice of Limited Enforcement Discretion for Existing Onshore Gas Transmission Pipelines (Dec. 6, 2022); Notice of Limited Enforcement Discretion for New and Replaced Onshore Gas Transmission Pipelines (April 17, 2023); Notice of Limited Enforcement Discretion for Certain Onshore Gas Transmission Pipelines Conducting Remedial Actions Under 49 CFR 192.473(c)(4) (Apr. 20, 2023). According to PHMSA, providing additional time can help facilitate more careful and comprehensive operator implementation efforts, and in some instances, for PHMSA to address preliminary implementation questions.

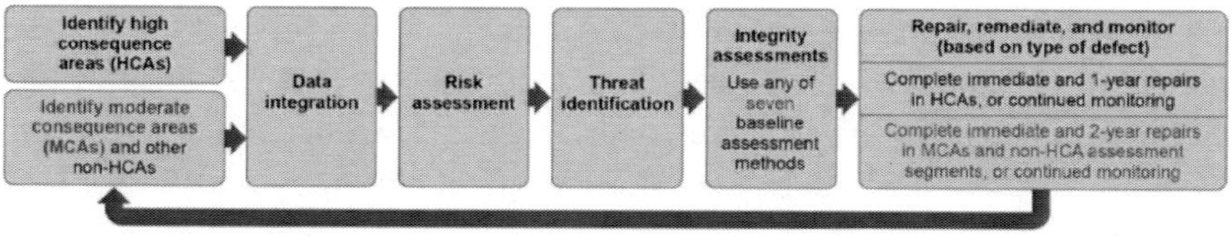

Source: GAO analysis of Pipeline amd Hazardous Materials Safety Adminitration (PHMSA) regulations; GAO (diagram). | GAO-24-106690.

Notes: See Pipeline Safety: Safety of Gas Transmission Pipelines: MAOP Reconfirmation, Expansion of Assessment Requirements, and Other Related Amendments, 84 Fed. Reg. 52180 (Oct. 1, 2019); Pipeline Safety: Safety of Gas Transmission Pipelines: Repair Criteria, Integrity Management Improvements, Cathodic Protection, Management of Change, and Other Related Amendments, 87 Fed. Reg. 52224 (Aug. 24, 2022).

While PHMSA's integrity assessment requirement applies to all gas transmission pipelines in HCAs, it applies only to these pipelines in MCAs that can accommodate in-line inspection tools, as well as such pipelines operating at or above 30 percent of the specified minimum yield strength in certain other non-HCAs.

Prior to the 2019 final rule, gas transmission pipeline operators could choose from three specific methods to conduct baseline integrity assessments of pipelines in HCAs and four specific methods for subsequent integrity assessments. Operators could also choose to use an alternative assessment method under certain conditions for both assessments.

Figure 4. Steps Operators Take to Assess Gas Transmission Pipelines, with Selected Regulatory Changes Made by PHMSA's 2019 and 2022 Final Rules.

What Methods Can Operators Use to Identify HCAs and MCAs on Their Pipelines?

Under integrity management regulations, pipeline operators can use two methods—class location and the potential impact radius (PIR)—to identify HCAs within their gas transmission pipeline system (see fig. 5). PHMSA's 2019 final rule required operators to use the PIR method to identify MCAs.

- *Class location.* A pipeline's class location is generally based on the population density within a specified distance of the pipeline's centerline. Using this method, operators generally count the number of buildings within 660 feet of a pipeline that are intended for human occupancy or occupied by a certain number of people for a specific amount of time. Since 1970, PHMSA has categorized the areas around gas pipelines into four classes. In general, Class 1 locations are in less populated areas, while Class 4 locations are in the most densely populated areas.

- *PIR.* The PIR method is also generally based on the population density of the area around the pipeline, but the size of that area is determined by calculating the individual pipeline's potential impact circle, rather than using a specific distance around the pipeline.[12] The PIR is the radius of this circle. The PIR is calculated using a regulatory formula that includes the diameter and maximum allowable operating pressure of the gas transmission pipeline as factors.

Which Methods Are Operators Using to Identify HCAs on Their Gas Transmission Pipelines?

Operators who reported complete data to PHMSA predominantly used the PIR method for most of their transmission pipeline mileage. PHMSA began collecting annual report data about the methods operators use to identify HCAs in 2022.

PHMSA's data show that operators used the PIR method for over 16,000 HCA transmission miles in 2022, compared to approximately 4,500 HCA miles with the class location method.

However, PHMSA's data on which method operators use are not complete. Our analysis of PHMSA's annual report data showed that nearly 40 percent of operators reported incomplete data on the methods they used in 2022. As a result, we are only reporting data for operators who submitted complete data to the agency. When we asked PHMSA officials about this issue, they told us that these inconsistencies were caused by an error in the agency's report submission system. PHMSA officials told us that the error was

---

[12] Under PHMSA's regulations governing the designation of HCAs using the PIR method, an HCA is an area within a potential impact circle containing either: (1) 20 or more buildings intended for human occupancy; or (2) an identified site. An identified site is: (1) an outside area or open structure that is occupied by 20 or more persons at a specific minimum time frequency; (2) a building that is occupied by 20 or more persons at a specific minimum time frequency; or (3) a facility occupied by persons who are confined or of impaired mobility or would be difficult to evacuate.

corrected in December 2023, before operators submitted their 2023 annual reports.

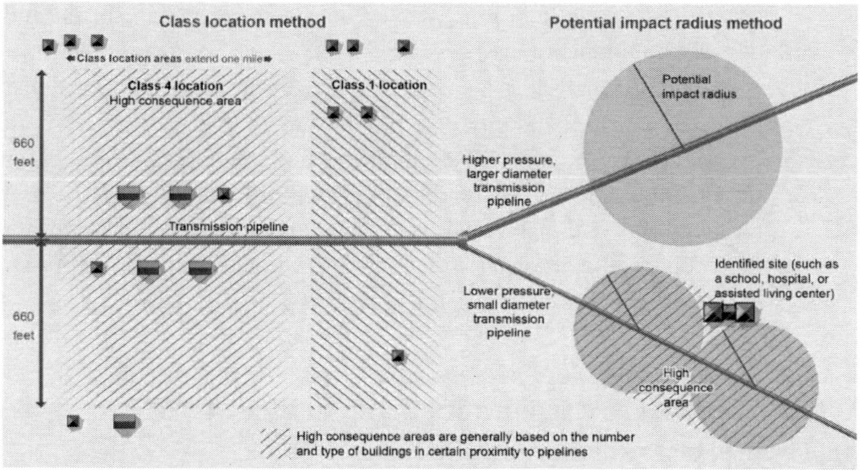

Source: GAO analysis of Pipeline Hazardous Materials Safety Administration regulations; GAO (illustration). GAO-24-106690.

Notes: These are examples of high consequence areas identified using each method. Under both methods, other areas may be identified as high consequence areas. An identified site is: (1) an outside area or open structure that is occupied by 20 or more persons at a specific minimum time frequency; (2) a building that is occupied by 20 or more persons at a specific minimum time frequency; or (3) a facility occupied by persons who are confined or of impaired mobility or would be difficult to evacuate. *See* 49 C.F.R. § 192.903.

Figure 5. Methods to Identify High Consequence Areas for Gas Transmission Pipelines.

We found that operators' choice of which method to use to identify HCAs generally varies by the type of pipeline. For instance, interstate operators or operators who have both interstate and intrastate pipelines tend to use the PIR method. These operators reported using the PIR method for over 7,500 HCA transmission miles and used the class location method for 615 HCA miles. When asked why they use the PIR method, one operator told us that it more accurately determines HCAs, because the method allows operators to more precisely identify higher risk pipeline segments. Another operator added that the PIR method helps focus resources on areas where they are most needed.

Operators who only have intrastate pipelines used the class location method more often than other operators. Intrastate operators who reported complete data in 2022 used the class location method for over 3,600 HCA pipeline miles, compared to 615 miles for interstate operators and those who have both interstate and intrastate pipelines. Operators of intrastate pipelines tend to be smaller (operate fewer pipeline miles) than interstate operators.

Three of 22 selected stakeholders told us that smaller operators, or those with fewer resources, may prefer to use the class location method over the PIR calculation. According to one operator, class location uses a simpler methodology and requires less data than using the PIR method. However, over half of the operators (six of 10) we spoke to said that the choice between the methods can be a strategic one. For instance, one operator said that using the class location method's more conservative approach fit better with the company's overall approach to safety.

What Changes Is PHMSA Considering to the Class Location and Potential Impact Radius Methods?

PHMSA proposed regulatory changes in 2020 to safety requirements for certain gas transmission pipeline segments when the class location of the pipeline has changed. Officials said PHMSA is also considering whether to initiate a rulemaking on changes to the PIR method, but we found the agency does not collect data that allows it to assess the accuracy of that method.

Class Location Method

PHMSA issued a notice of proposed rulemaking to amend the class location change regulations in 2020. Among other changes, the proposed rule would offer operators an alternative for managing class location changes that is based on integrity management.[13] Specifically, this alternative would be available when an eligible gas transmission pipeline segment changes from a less populated Class 1 location to a more populated Class 3 location. The proposed rule would also modify the definition of an HCA to include these pipeline segments, so that integrity management program requirements would apply to them. Under the existing regulations, operators must reduce a gas transmission pipeline's operating pressure, replace the pipe, or pressure test to higher standards when the class location around it increases. PHMSA officials told

[13] Pipeline Safety: Class Location Change Requirements, 85 Fed. Reg. 65142 (Oct. 14, 2020).

us that the proposed rule is scheduled to be discussed with stakeholders at the March 2024 meeting of the Gas Pipeline Advisory Committee.

PIR Method

According to PHMSA officials, PHMSA is also considering changes to the PIR calculation in response to a recent National Transportation Safety Board (NTSB) recommendation. In 2022, NTSB found that PHMSA's equation for determining the PIR of a pipeline rupture does not match findings from investigations of recent gas ruptures and is based on assumptions that may not be realistic. For example, NTSB found that the equation assumes that an individual near a rupture would be able to evacuate or find shelter in 30 seconds. This assumption may not be accurate for individuals with mobility impairments or pre-existing medical conditions. Since NTSB's findings, PHMSA held a public meeting with a series of presentations on the PIR in December 2022, seeking input from industry and public stakeholders. In October 2023, PHMSA officials told us that the agency established a team to review the PIR calculation after the public meeting, and that PHMSA is evaluating whether changes may be needed to the PIR.

PHMSA does not collect detailed data comparing the PIR calculation to the actual impact of an incident on that pipeline, because the agency's incident form is not designed to collect that information. For each of the three questions on the form related to damage occurring outside of a pipeline's PIR, PHMSA asks operators to indicate if damage occurred via a yes/no response; however, it does not ask operators how far outside the PIR any damage occurred. As a result, PHMSA cannot analyze these incident characteristics in its database.

In addition, operator incident reports do not always provide detailed explanations for any injuries and fatalities that occur outside of the pipeline's PIR. For example, NTSB's 2022 report described how and where a fatality occurred outside of a Kentucky pipeline's calculated potential impact circle. In the incident report submitted to PHMSA, the operator indicated that the fatality occurred but did not include many of these details.

PHMSA officials told us that the agency typically conducts safety investigations following reportable incidents to ensure that operators determine the root cause of the failure and are operating in compliance with all regulations. Officials said the incident report forms are used as a catalyst for initiating those investigations.

According to Standards for Internal Control in the Federal Government, to best achieve their objectives, agencies should obtain relevant data from reliable external sources.[14] As PHMSA is considering changes to the PIR calculation, determining what additional data are needed from operators could improve the quality of PHMSA's initial data on actual pipeline incidents and assist the agency in evaluating operator methods to determine HCAs over time.

Without incident forms that allow PHMSA to assess the full impact of certain pipeline incidents, the agency may find it more difficult to assess whether a particular method is accurate for operators identifying HCAs. Having such information would provide PHMSA with additional data it could use as it considers whether to propose regulatory changes to the PIR equation.

Which Changes in the 2019 And 2022 Final Rules Do Stakeholders Think Will Have The Largest Impact on Pipeline Safety?

Selected stakeholders and officials from PHMSA regional offices said that they expect several key changes in the 2019 and 2022 final rules will improve pipeline safety by increasing the miles of pipeline assessed, improving operator data and information, and strengthening preventative and remedial requirements for operators.[15] These stakeholders included representatives from pipeline industry and safety groups, state regulatory officials, and gas transmission pipeline operators. In addition, given the long time frames involved with the rules, some stakeholders (eight of 22) and officials from several PHMSA regional offices (three of five) noted that it may be too early to estimate the extent of safety benefits.[16]

Stakeholders and PHMSA regional officials identified the following key changes from the 2019 rule:

---

[14] GAO, Standards for Internal Control in the Federal Government, GAO-14-704G (Washington, D.C.: Sept. 10, 2014).

[15] Some of these key changes apply to gas transmission pipelines subject to integrity management program requirements, while most affect gas transmission pipelines not subject to these requirements, such as those located in MCAs.

[16] We conducted semi-structured interviews with stakeholders, and not all interviewees discussed certain topics. In our analysis of interview responses, we analyzed the number of interviewees that made a given statement; as such, interviewees who did not comment on a particular topic did not necessarily disagree.

Increasing Miles of Pipeline Assessed

Assessing Areas Outside of HCAs (17 of 22 stakeholders, four of five PHMSA regional offices). The rule changes included establishing the MCA definition and requiring operators to conduct integrity assessments of certain gas transmission pipelines in MCAs and specific class locations using the same assessment methods adopted for HCAs.[17] In 2022, PHMSA's data show that this requirement added an additional 24,000 miles to operators' total assessed mileage, increasing the total assessed mileage by 115 percent to nearly 46,000 miles. Five operators said they were already assessing mileage beyond HCAs on their own before the final rule established the requirement to assess MCA and certain class location mileage; they said requiring all operators to do so was beneficial. One operator said that expanding the integrity assessment requirements beyond HCAs requires operators to know more about their pipelines; another said that having greater knowledge of their pipeline system makes it safer.

Improving Data and Information

- Maximum Allowable Operating Pressure Reconfirmation (15 of 22 stakeholders, four of five PHMSA regional offices). The rule required operators to reconfirm their pipelines' maximum allowable operating pressure under certain conditions; it provided six methods that operators could use to do so. Affected pipelines include those in HCAs and certain pipelines in Class 3 or 4 locations that have not previously been subject to a pressure test or that lack the traceable, verifiable, and complete records needed to establish the pipeline's maximum allowable operating pressure. An operator said that having additional methods to reconfirm the maximum allowable operating pressure is helpful if records are not complete, because it gives them more options to test their pipelines, rather than having to replace the pipe.
- Verification of Pipeline Material Properties and Attributes (11 of 22 stakeholders, three of five PHMSA regional offices). The rule

[17] Specifically, the 2019 final rule requires integrity assessments of gas transmission pipelines that operate at a certain pressure and: (1) are located in an MCA and able to accommodate inspection using a certain tool; or (2) located in a Class 3 or 4 location.

established a requirement for operators to review the property and material records for steel pipelines to ensure the records are traceable, verifiable, and complete. Two industry stakeholders said this change was important because it required operators to ensure that their records matched the pipe in the ground. One operator said this change was beneficial for pipeline safety, because the company had recently acquired pipelines with poor records.

Stakeholders and PHMSA regional officials identified the following key changes from the 2022 rule:

Strengthening Preventative and Remedial Requirements

- *Repair Criteria (10 of 22 stakeholders, four of five PHMSA regional offices).* The rule required immediate repairs of additional pipeline defects in HCAs, which an operator must repair upon discovery. Defects could include a dent, crack, or certain types of metal loss. The rule also establishes repair criteria for defects found in MCAs and other non-HCA areas, which will require certain repairs within 2 years of discovery or continued monitoring. One stakeholder said this change is beneficial because it expands the number of miles of pipeline subject to repair criteria and requires operators to address issues sooner.
- *Corrosion Control (10 of 22 stakeholders).* The rule strengthened regulations for corrosion control with new requirements for pipe coating assessments and protective coating strength, among other things. One stakeholder told us this was a critical change because it prescribes the steps operators must take to comply with the rule.

What Challenges Have Pipeline Operators and Inspectors Faced in Applying the Final Rules?

Some selected stakeholders said that the timeliness and clarity of PHMSA's guidance for the 2019 final rule posed a challenge for operators and inspectors applying the final rule changes (i.e., complying with the final rules and evaluating operator compliance). In addition, some stakeholders said that allocating resources to comply with the 2019 final rule was challenging due to the additional time and resources needed to make the changes, which stakeholders said PHMSA underestimated when it developed the rule. PHMSA postponed enforcing the 2022 final rule until February 2024 to help improve operator compliance and implementation efforts, according to PHMSA officials.

Timeliness and Clarity of Guidance

- Timely and Clear FAQs (12 of 22 stakeholders). Over half of the stakeholders said the FAQs PHMSA issued for the 2019 final rule were not timely or clear.[18] For example, PHMSA published the first set of FAQs 2 months after the 2019 final rule went into effect and published the second set of FAQs more than 2 years later. An operator said that it was difficult to understand which pipeline segments fell under specific regulations, for example, which pipelines were required to get material verification or reconfirm the maximum allowable operating pressure; the FAQs did not provide further clarity.

  According to PHMSA officials, the timing of FAQs was affected by the length of the comment period for the draft FAQs and operational delays from the COVID-19 pandemic. In addition, they noted that FAQs explain existing regulations but do not further restrict operators or alter the rules they are based on.
- Consistency in Evaluating Final Rule Compliance (nine of 22 stakeholders). Some stakeholders noted a need for consistency between federal and state inspectors when applying changes to the regulations. One operator said that a lack of consistency among inspectors has led to confusion for operators on how to comply with the changes made by the 2019 final rule. According to one official from a state pipeline agency, state inspectors also experience challenges during inspections. The official estimated that some inspectors can spend hours trying to figure out which regulations apply during an inspection and specifically noted a lack of clarity around operating pressure records requirements.

---

[18] PHMSA issues FAQs to assist pipeline owners and operators in complying with regulatory changes.

- PHMSA officials told us that consistency in applying the changes to the regulations is a top priority. According to officials, PHMSA established a small team with federal and state inspectors for several initial inspections to test and retool their inspection approach to help ensure a more consistent approach for the 2019 rule. PHMSA officials said they will continue this approach for the 2022 final rule.

Resource Allocation

- *Additional time and resources (11 of 22 stakeholders).* Half of the selected stakeholders said the new requirements established by the final rules will require additional time and resources for operators, such as additional personnel. For example, one operator said that the compliance deadlines for the rule have been difficult for operators to meet due to the large-scale changes they needed to make to their operations. However, the operator also noted that PHMSA postponed enforcement of the 2022 final rule to give operators more time. In addition, a smaller operator said that the company did not have the resources for in-house experts on integrity management and would need to hire outside experts for guidance on compliance with the changes in the final rules.
- Underestimated Costs (seven of 22 stakeholders). Some stakeholders said PHMSA's cost-benefit analysis in the regulatory impact assessments for the final rules did not capture the full extent of costs for operators. One operator said that the assessments made simplistic assumptions for a highly complex rule.

What Steps Has PHMSA Taken to Implement the 2019 and 2022 Final Rules?

PHMSA has largely implemented the 2019 final rule. PHMSA officials said they are in the process of implementing the 2022 final rule, but they have not fully documented their activities by creating an implementation plan.

2019 Final Rule Implementation Activities

Shortly after issuing the 2019 final rule, PHMSA officials said they formed a temporary implementation team to carry out activities, such as holding workshops with state and federal inspectors and issuing guidance to communicate the rule changes to inspectors and the pipeline industry. According to PHMSA officials, these activities were designed to prepare inspectors to conduct assessments and help operators understand how to comply with the final rule changes.

PHMSA performed the following activities to implement the 2019 final rule:

- Hosted a public meeting covering early rule implementation activities in February 2020 for industry, state agencies, and the general public;
- Held in-person trainings and workshops and hosted virtual webinars for federal and state inspectors from 2020-2023, according to PHMSA officials;
- Issued the first set of FAQs in September 2020 and the second set in April 2023;
- Issued a public inspection question set in July 2021, intended to educate PHMSA and state pipeline safety inspectors on pipeline safety standards amended by the final rule, and according to PHMSA officials, conducted pilot and post-pilot inspections;
- Hosted a Gas Pipeline Advisory Committee meeting for industry in October 2021 to provide an update on the implementation of the rule.

In 2021, while PHMSA was implementing the 2019 final rule, PHMSA's Office of Program Development created the Fundamental Guide for Pipeline Safety Final Rule Implementation (Implementation Guide), an internal document that provides guidelines for implementing final rules. During this time, PHMSA's temporary implementation team went through organizational changes, and PHMSA officials could not provide an implementation plan for the 2019 final rule. However, PHMSA officials provided a task list documenting the key activities the agency completed to implement the 2019 final rule. As of February 2024, PHMSA has largely finished implementing the 2019 final rule.

2022 Final Rule Implementation Activities

According to PHMSA officials, in 2022 the agency formed a new implementation division to implement the 2022 final rule and future rules. Officials said that PHMSA recognized a need for a dedicated team to facilitate the implementation of rules that are increasingly complex. In January 2024, officials said that this division was in the process of implementing the 2022 final rule and following an updated draft version of the Implementation Guide but has not created an implementation plan for the 2022 rule.[19]

According to Standards for Internal Control in the Federal Government, agencies should internally and externally communicate the necessary quality information to achieve their objectives.[20] Agencies should also consider factors including the purpose of the information, the intended recipients of the information, and when the audience needs to receive the information. PHMSA includes these considerations in its Implementation Guide. The updated version of the Implementation Guide states that a team implementing a final rule should develop an implementation plan that communicates, among other things, what the implementation team intends to achieve, time frames when deliverables will be complete, and a communication strategy to relay information to the right audience.

PHMSA officials said they do not intend to develop an implementation plan as described in the Implementation Guide, because the Guide was being updated at the time and some implementation tasks, such as one set of FAQs, have already been completed.

Instead of an implementation plan, officials created an internal task list to document their implementation activities and keep track of timelines. However, our review of PHMSA's task list for the 2022 final rule found that the list was a summary of actions, rather than an implementation plan with clear objectives, time frames, and an outreach strategy. The task list documents certain milestone dates such as the publication of the final rule and its effective dates, but not timelines for when other implementation activities, such as updates to PHMSA's enforcement manual, must be accomplished. In addition, the task list does not include the purpose of the activities, the intended audience for each activity, or how PHMSA will communicate with that audience.

---

[19] PHMSA's Office of Pipeline Safety issued an updated version of the Rule Implementation Guide in February 2024.

[20] GAO-14-704G.

PHMSA's remaining activities for the 2022 final rule include inspection questions and training for federal and state inspectors, according to PHMSA officials.

PHMSA officials said they will not implement these activities until February 2024, when the agency will begin enforcing all provisions of the 2022 final rule. PHMSA also intends to issue other sets of FAQs for the 2022 final rule, among other activities. Officials said some of their activities have been delayed due to legal challenges and PHMSA's notices of enforcement discretion.[21] For example, PHMSA explained that they waited to issue the first set of FAQs for the rule until January 2024, in part due to ongoing litigation in federal court challenging certain provisions of the 2022 final rule.[22] According to PHMSA officials, they do not have formal requirements to issue FAQs, but officials acknowledged that they have become expected by the industry in light of the increasing complexity of pipeline safety regulations.[23]

However, the remaining activities are key to helping stakeholders understand the requirements of the 2022 final rule. As implementation activities are still ongoing, the agency has an opportunity to develop an implementation plan that could help ensure it accomplishes the remaining activities and conducts outreach to the appropriate audiences in a timely manner. With a robust implementation plan, PHMSA would be in a better position to provide relevant guidance and information to operators and inspectors.

---

[21] These challenges included joint petitions for reconsideration of certain provisions of the 2022 final rule. *See, e.g.*, Petition for Reconsideration of the Interstate Natural Gas Association of America and the American Petroleum Institute, Doc. No. PHMSA-2011-0023-0644 (received by PHMSA on Oct. 20, 2022). PHMSA and the petitioners met over several months to try to resolve the issues raised in these petitions, according to both PHMSA officials and industry association representatives.

[22] In July 2023, the Interstate Natural Gas Association of America (INGAA) filed a lawsuit challenging parts of the final rule. *See* Petition for Review, INGAA v. PHMSA, Docket No. 23-1173 (D.C. Cir.) (filed on June 10, 2023). According to PHMSA officials, INGAA made clear the specific provisions that it was disputing in its brief filed on December 5, 2023, and as a result, the agency issued the FAQs only on the undisputed provisions of the 2022 final rule. PHMSA officials told us that they did not want to issue guidance on the disputed provisions of the rule, as it might need to be amended or retracted depending on the outcome of the litigation.

[23] According to PHMSA, several factors impact the development and publication of FAQs, including the complexity of the regulation, any ongoing litigation challenging the final rule, public comments on the final rule or questions from operators, and compliance issues that are identified during state and federal inspections.

To What Extent Do PHMSA's Web Pages Provide Complete Information About the 2019 and 2022 Final Rules?

In 2023, PHMSA created rule implementation web pages with information about the 2019 and 2022 final rules. However, PHMSA's resources and information relevant to rule implementation and compliance are not always linked to PHMSA's rule implementation web pages, which officials told us are meant to provide easy-to-find information and guidance on the recent rulemakings.

We found that the web pages for the 2019 and 2022 final rules do not include the full scope of PHMSA's outreach activities. Information about the final rules is in several places on PHMSA's website, where it may not be easily accessible to operators or state inspectors. PHMSA has a guidance portal for operators on its website, in addition to the rule implementation web pages. PHMSA officials said the portal is designed to serve as a resource for industry and the public. The portal provides a variety of compliance resources, guidance manuals, and FAQs, among other resources. However, the portal does not help operators identify additional presentations or resources on the 2019 and 2022 final rules not linked on PHMSA's implementation web pages. For example, to find the industry meeting where PHMSA officials discussed the implementation of the 2019 final rule and the results of PHMSA's pilot inspections, a user would need to navigate through the agency's Meetings and Events web page and know the date of the meeting that covered the 2019 final rule. Alternatively, the user would need to download and review the agendas for each meeting.

According to Standards for Internal Control in the Federal Government, agencies should communicate externally the necessary quality information to help achieve their objectives.[24] In addition, PHMSA's Implementation Guide states that developing an outreach strategy that provides clear information to a targeted audience is a key component of an implementation plan. When asked about the goal and intended audience of the Final Rule Implementation web pages, PHMSA officials stated that the web pages are intended to provide complete information on the recent final rules for anyone interested. In addition, the Pipeline Rulemaking Implementation homepage states that the "information is oriented primarily toward operators to provide information useful for complying with the pipeline safety regulations." Having complete information on these web pages would better ensure that they are useful for

[24] GAO-14-704G.

gas transmission operators trying to comply with PHMSA's pipeline safety regulations.

Without action from PHMSA to update its Final Rule Implementation web pages, operators and inspectors may have difficulties when trying to find information on the 2019 and 2022 final rules.

Conclusion

If PHMSA decides to propose regulatory changes to the PIR calculation, it could affect how operators identify HCAs and MCAs in the future. Without specific data comparing the calculated potential impact of a pipeline incident to the actual impact that occurred, PHMSA will not be able to fully evaluate how the PIR calculation could be changed to more accurately predict the impact of incidents and improve safety outcomes for the people involved. Assessing what additional data are needed from operator incident reports could improve the quality of PHMSA's data on actual pipeline incidents to help strengthen its analysis as it considers potential regulatory changes.

Stakeholders and PHMSA regional officials generally concurred that PHMSA's 2019 and 2022 final rules hold promise to improve the safety of gas transmission pipelines by strengthening operator requirements and extending assessment requirements to pipelines beyond HCAs. While the agency has largely implemented the 2019 final rule, developing a plan for the 2022 rule would help ensure that PHMSA accomplishes its remaining planned activities and conducts outreach to the appropriate audiences in a timely manner. Further, having complete information on the 2019 and 2022 final rules on PHMSA's Final Rule Implementation web pages would better ensure that these web pages are useful for operators and inspectors.

Recommendations for Executive Action

We are making the following three recommendations to PHMSA:

- The Administrator of PHMSA should, as PHMSA considers possible changes to the potential impact radius calculation, evaluate what additional data are needed from operators to better understand the actual impact of pipeline incidents. (Recommendation 1)

- The Administrator of PHMSA should develop an implementation plan for the remaining activities for the 2022 final rule that includes clear objectives, timelines, and an outreach strategy. (Recommendation 2)
- The Administrator of PHMSA should update the 2019 and 2022 Gas Transmission Final Rule Implementation web pages to increase accessibility to rule implementation information. (Recommendation 3)

Agency Comments

We provided a draft of this report to the Department of Transportation for review and comment. In its comments, reproduced in appendix I, the agency concurred with our recommendations. The Department of Transportation also provided technical comments, which we incorporated as appropriate.

How GAO Did This Study

To examine the methods operators use to identify their HCAs and assess the extent to which PHMSA is planning any changes to the methods, we reviewed PHMSA's regulations governing these methods and interviewed PHMSA officials. We analyzed PHMSA's gas transmission pipeline incident data from 2010 through 2022, which includes data on the actual impact and characteristics of pipeline incidents. In addition, we analyzed PHMSA's 2022 gas transmission pipeline annual report data submitted in 2023, including data on HCA mileage and the methods operators used to identify their HCAs.

We assessed data reliability using several methods, including reviewing related documentation, interviewing officials, and conducting several data checks. During our analysis, we found that nearly 40 percent of operator reports submitted inconsistent annual report data when reporting HCA mileage by method.[25] To account for this limitation, we acknowledged these inconsistences and only reported complete data. PHMSA officials told us they

[25] PHMSA considers data in this field to be consistent if (1) operators with over 50 HCA miles report mileage by HCA method within 0.2 percent of total HCA mileage; or (2) operators with less than 50 HCA miles report mileage by HCA method within 0.1 miles of total HCA mileage.

planned to correct this error in their system before operators began submitting 2023 pipeline data. Outside of this limitation, we found both datasets to be sufficiently reliable for the purposes of our report.

To identify key regulatory changes that stakeholders believe will impact gas transmission pipeline safety, we collected a range of perspectives by interviewing a non-generalizable sample of four industry associations, three pipeline safety stakeholders, five state pipeline safety agencies, and 10 gas transmission pipeline operators identified through PHMSA's 2022 annual report data.[26] For the state agencies, we selected one state from each PHMSA geographic region— Central, Eastern, Southern, Southwestern, and Western. We selected states that had a significant amount of transmission pipeline mileage, along with variation in (1) whether the state assumed jurisdiction over the intrastate or interstate pipelines within its borders, and (2) the number of gas transmission pipeline incidents in the state from January 2018 through May 2023. In each selected state, we interviewed officials from the designated gas pipeline safety agency and two pipeline operators that were randomly selected to allow for variation in the amount of HCA pipeline mileage they reported to PHMSA, whether the operator had multiple pipeline commodities, and whether they operated intra- or interstate pipelines.[27] One of the operators provided written responses in lieu of an interview.

To examine PHMSA's recent regulatory changes to gas transmission pipeline safety and assess the extent to which PHMSA is overseeing the implementation of the 2019 and 2022 rules, we reviewed relevant statutes, regulations, and PHMSA rulemakings and documents. These documents included the 2019 and 2022 final gas transmission pipeline safety rules, PHMSA's guidance documents related to the rules and rule implementation,

---

[26] The industry associations we interviewed were the American Gas Association, American Public Gas Association, Interstate Natural Gas Association of America, and National Association of Pipeline Safety Representatives. The pipeline safety stakeholders we interviewed were the National Transportation Safety Board, Pipeline Safety Trust, and a pipeline safety consultant.

[27] The 10 pipeline operators we interviewed were DTE Gas Company, Energy Transfer Company, Empire Pipeline, Enmark Energy, Gulf South Pipeline Company/Boardwalk Pipelines, Michigan Gas Utilities, North Baja Pipeline/TC Energy, Seadrift Pipeline Corp/Dow Pipeline Company, Southern California Gas Company, and Valley Energy. The state pipeline agencies selected for an interview were the California Public Utilities Commission, Michigan Public Service Commission, Mississippi Public Service Commission, Pennsylvania Public Utility Commission, and the Railroad Commission of Texas.

and PHMSA's annual report and incident forms. We also reviewed PHMSA documentation on the agency's implementation activities related to the final rules and interviewed agency officials and officials from the five PHMSA regional offices about these activities. We compared PHMSA's activities to criteria identified in Standards for Internal Controls in the Federal Government.[28]

We conducted this performance audit from March 2023 to April 2024 in accordance with generally accepted government auditing standards. Those standards require that we plan and perform the audit to obtain sufficient, appropriate evidence to provide a reasonable basis for our findings and conclusions based on our audit objectives. We believe that the evidence obtained provides a reasonable basis for our findings and conclusions based on our audit objectives.

List of Addressees

The Honorable Maria Cantwell
Chair

The Honorable Ted Cruz
Ranking Member
Committee on Commerce, Science, and Transportation
United States Senate

The Honorable Cathy McMorris Rodgers
Chair
The Honorable Frank Pallone, Jr.
Ranking Member
Committee on Energy and Commerce
House of Representatives

The Honorable Sam Graves
Chairman

[28] GAO-14-704G.

The Honorable Rick Larsen Ranking
Member
Committee on Transportation and Infrastructure
House of Representatives

We are sending copies of this report to the appropriate congressional committees, the Secretary of Transportation, and other interested parties. In addition, the report is available at no charge on the GAO web site at https://www.gao.gov.

GAO Contact Information

For more information, contact: Elizabeth Repko, Director, Physical Infrastructure, RepkoE@gao.gov, (202) 512-2834.

Chuck Young, Managing Director, Public Affairs, YoungC1@gao.gov, (202) 512- 4800.

A. Nicole Clowers, Managing Director, Congressional Relations, ClowersA@gao.gov, (202) 512-4400.

Staff Acknowledgments

Sara Vermillion (Assistant Director), Melissa Greenaway (Analyst-in-Charge), Madeline Barch, Joshua Ormond, Mary Catherine P. Overcash, Kristin Petroff, Jack Wang, and Amelia Michelle Weathers.

Connect with GAO on Facebook, Flickr, Twitter, and YouTube. Subscribe to our RSS Feeds or Email Updates. Listen to our Podcasts.

Visit GAO on the web at https://www.gao.gov.

Appendix I: Comments from Department of Transportation

U.S. Department of Transportation
Office of the Secretary of Transportation

Assistant Secretary for Administration

1200 New Jersey Avenue, SE
Washington, DC 20590

March 18, 2024

Ms. Elizabeth Repko
Director, Physical Infrastructure
U.S. Government Accountability Office
441 G Street NW
Washington, DC 20548

Dear Ms. Repko:

The mission of the Pipeline and Hazardous Materials Safety Administration (PHMSA) is to protect people and the environment by advancing the safe transportation of energy and other hazardous materials that are essential to our daily lives. To do this, the agency establishes national policy, sets and enforces standards, educates, and conducts research to prevent incidents. As a part of its mission, PHMSA has responsibility for ensuring the safety of the nation's pipeline system in partnership with other state and Federal agencies and pipeline owners and operators. PHMSA will resolve any outstanding issues on the implementation plan for the 2022 final rule "Pipeline Safety: Safety of Gas Transmission Pipelines: Repair Criteria, Integrity Management Improvements, Cathodic Protection, Management of Change, and Other Related Amendments" once litigation concerning the final rule is resolved. In addition, PHMSA is evaluating available information regarding modification of existing regulations for calculating potential impact radius and continues to update our webpages to improve accessibility.

Upon review of the draft report, PHMSA concurs with the Government Accountability Office's (GAO's) recommendation to evaluate what additional data is needed from operators to better understand the actual impact of pipeline incidents as PHMSA considers updating the potential impact radius calculation, and the recommendation to update the 2019 and 2022 Gas Transmission Final Rule Implementation webpages to increase accessibility to rule implementation information. PHMSA also concurs with the recommendation to develop an implementation plan for the remaining activities for the 2022 final rule that includes clear objectives, timelines, and an outreach strategy. PHMSA appreciates the opportunity to respond to the GAO draft report and will provide a detailed response to each recommendation within 180 days of the final report's issuance.

Please contact Gary Middleton, Director of Audit Relations and Program Improvement at (202) 366-6512 or gary.middleton@dot.gov with any questions or if GAO would like additional information.

Sincerely,

Philip McNamara
Assistant Secretary for Administration

Index

R

S

T

U

V

W